AUTOMATIC PROGRAMMING
APPLIED TO VLSI CAD SOFTWARE:
A Case Study

THE KLUWER INTERNATIONAL SERIES IN ENGINEERING AND COMPUTER SCIENCE
VLSI, COMPUTER ARCHITECTURE AND DIGITAL SIGNAL PROCESSING
Consulting Editor
Jonathan Allen

Other books in the series:

Adaptive Filters: Structures, Algorithms, and Applications. M.L. Honig and D.G. Messerschmitt. ISBN 0-89838-163-0.
Introduction to VLSI Silicon Devices: Physics, Technology and Characterization. B. El-Kareh and R.J. Bombard. ISBN 0-89838-210-6.
Latchup in CMOS Technology: The Problem and Its Cure. R.R. Troutman. ISBN 0-89838-215-7.
Digital CMOS Circuit Design. M. Annaratone. ISBN 0-89838-224-6.
The Bounding Approach to VLSI Circuit Simulation. C.A. Zukowski. ISBN 0-89838-176-2.
Multi-Level Simulation for VLSI Design. D.D. Hill and D.R. Coelho. ISBN 0-89838-184-3.
Relaxation Techniques for the Simulation of VLSI Circuits. J. White and A. Sangiovanni-Vincentelli. ISBN 0-89838-186-X.
VLSI CAD Tools and Applications. W. Fichtner and M. Morf, Editors. ISBN 0-89838-193-2.
A VLSI Architecture for Concurrent Data Structures. W.J. Dally. ISBN 0-89838-235-1.
Yield Simulation for Integrated Circuits. D.M.H. Walker. ISBN 0-89838-244-0.
VLSI Specification, Verification and Synthesis. G. Birtwistle and P.A. Subrahmanyam. ISBN 0-89838-246-7.
Fundamentals of Computer-Aided Circuit Simulation. W.J. McCalla. ISBN 0-89838-248-3.
Serial Data Computation. S.G. Smith, P.B. Denyer. ISBN 0-89838-253-X.
Phonologic Parsing in Speech Recognition. K.W. Church. ISBN 0-89838-250-5.
Simulated Annealing for VLSI Design. D.F. Wong, H.W. Leong, C.L. Liu. ISBN 0-89838-256-4.
Polycrystalline Silicon for Integrated Circuit Applications. T. Kamins. ISBN 0-89838-259-9.
FET Modeling for Circuit Simulation. D. Divekar. ISBN 0-89838-264-5.
VLSI Placement and Global Routing Using Simulated Annealing. C. Sechen. ISBN 0-89838-281-5.
Adaptive Filters and Equalizers. B. Mulgrew, C.F.N. Cowan. ISBN 0-89838-285-8.
Computer-Aided Design and VLSI Device Development, Second Edition. K.M. Cham, S-Y. Oh, J.L. Moll, K. Lee, P. Vande Voorde, D. Chin. ISBN: 0-89838-277-7.
Automatic Speech Recognition. K-F. Lee. ISBN 0-89838-296-3.
Speech Time-Frequency Representations. M.D. Riley. ISBN 0-89838-298-X.
A Systolic Array Optimizing Compiler. M.S. Lam. ISBN: 0-89838-300-5.
Algorithms and Techniques for VLSI Layout Synthesis. D. Hill, D. Shugard, J. Fishburn, K. Keutzer. ISBN: 0-89838-301-3.
Switch-Level Timing Simulation of MOS VLSI Circuits. V.B. Rao, D.V. Overhauser, T.N. Trick, I.N. Hajj. ISBN 0-89838-302-1.
VLSI for Artificial Intelligence. J.G. Delgado-Frias, W.R. Moore (Editors). ISBN 0-7923-9000-8.
Wafer Level Integrated Systems: Implementation Issues. S.K. Tewksbury. ISBN 0-7923-9006-7.
The Annealing Algorithm. R.H.J.M. Otten & L.P.P.P. van Ginneken. ISBN 0-7923-9022-9.
VHDL: Hardware Description and Design. R. Lipsett, C. Schaefer and C. Ussery. ISBN 0-7923-9030-X.
The VHDL Handbook. D. Coelho. ISBN 0-7923-9031-8.
Unified Methods for VLSI Simulation and Test Generation. K.T. Cheng and V.D. Agrawal. ISBN 0-7923-9025-3.
ASIC System Design with VHDL: A Paradigm. S.S. Leung and M.A. Shanblatt. ISBN 0-7923-9032-6.
BiCMOS Technology and Applications. A.R. Alvarez (Editor). ISBN 0-7923-9033-4.
Analog VLSI Implementation of Neural Systems. C. Mead and M. Ismail (Editors). ISBN 0-7923-9040-7.
The MIPS-X RISC Microprocessor. P. Chow. ISBN 0-7923-9045-8.
Nonlinear Digital Filters: Principles and Applications. I. Pitas and A.N. Venetsanopoulos. ISBN 0-7923-9049-0.
Algorithmic and Register-Transfer Level Synthesis: The System Architect's Workbench. D.E. Thomas, E.D. Lagnese, R.A. Walker, J.A. Nestor, J.V. Rajan, R.L. Blackburn. ISBN 0-7923-9053-9.
VLSI Design for Manufacturing: Yield Enhancement. S.W. Director, W. Maly, A.J. Strojwas. ISBN 0-7923-9053-7.
Testing and Reliable Design of CMOS Circuits. N.K. Jha, S. Kundu. ISBN 0-7923-9056-3.
Hierarchical Modeling for VLSI Circuit Testing. D. Bhattacharya, J.P. Hayes. ISBN 0-7923-9058-X.
Steady-State Methods for Simulating Analog and Microwave Circuits. K. Kundert, A. Sangiovanni-Vincentelli, J. White. ISBN 0-7923-9069-5.
Introduction to Analog VLSI Design Automation. M. Ismail, J. Franca. ISBN 0-7923-9071-7.
Principles of VLSI System Planning: A Framework for Conceptual Design. A.M. Dewey, S.W. Director. ISBN 0-7923-9102-0.
Mixed-Mode Simulation. R. Saleh, A.R. Newton. ISBN 0-7923-9107-1.

AUTOMATIC PROGRAMMING APPLIED TO VLSI CAD SOFTWARE:
A Case Study

by

Dorothy E. Setliff
University of Pittsburgh

and

Rob A. Rutenbar
Carnegie Mellon University

KLUWER ACADEMIC PUBLISHERS
Boston/Dordrecht/London

Distributors for North America:
Kluwer Academic Publishers
101 Philip Drive
Assinippi Park
Norwell, Massachusetts 02061 USA

Distributors for all other countries:
Kluwer Academic Publishers Group
Distribution Centre
Post Office Box 322
3300 AH Dordrecht, THE NETHERLANDS

Library of Congress Cataloging-in-Publication Data

Setliff, Dorothy E., 1958-
    Automatic programming applied to VLSI CAD software : a case study
/ by Dorothy E. Setliff and Rob A. Rutenbar.
      p. cm.
    Includes bibliographical references.
    ISBN 0-7923-9112-8
    1. Automatic programming (Computer science) 2. Computer-aided
design. 3. Integrated circuits—large scale integration.
I. Rutenbar, Rob A., 1957-   . II. Title.
QA76.6.S468 1990
620 '.0042 '0285—dc20   90-34690
                                                                         CIP

Copyright © 1990 by Kluwer Academic Publishers

All rights reserved. No part of this publication may be reproduced, stored in a retrieval system or transmitted in any form or by any means, mechanical, photocopying, recording, or otherwise, without the prior written permission of the publisher, Kluwer Academic Publishers, 101 Philip Drive, Assinippi Park, Norwell, Massachusetts 02061.

Printed in the United States of America

Table of Contents

1. Introduction — 1
1.1. The Application Domain — 4
1.2. Knowledge Sources — 6
1.3. Book Organization — 8
References — 11
2. Application Domain: Routing Algorithms — 13
2.1. Routing Algorithms — 13
2.1.1. Router Classifications — 14
2.1.2. Types of Routing Constraints — 18
2.2. Application Domain: Maze Routers — 19
2.3. Maze Router Varieties — 24
2.3.1. Fabrication Constraints — 25
2.3.2. Application Constraints — 26
2.3.3. Algorithm Constraints — 27
2.4. Why Choose Maze Routers? — 29
2.5. Chapter Summary — 30
References — 31
3. Software Reusability — 33
3.1. Composition-Based Systems — 34
3.2. Generation-Based Systems — 35
3.3. Chapter Summary — 39
References — 41
4. ELF: A Program Synthesis Architecture — 43
4.1. Combining Router Knowledge with Program Synthesis Knowledge — 44
4.2. Algorithm Schema Representation — 47
4.3. Data Structure Style Representation — 50
4.4. Intermediate Representation for Synthesized Code — 53
4.5. Domain Knowledge Representation Using a Rule-Based System — 54
4.5.1. Design Generation Knowledge — 56
4.5.1.1. Router Structure Knowledge — 56
4.5.1.2. Routing Phase Requirement Knowledge — 57
4.5.1.3. Router Dependency Knowledge — 57
4.5.2. Program Synthesis Knowledge — 58
4.5.2.1. Application Language Syntactic Knowledge — 59
4.5.2.2. Data Structure Implementation Knowledge — 59
4.5.3. Domain Interaction Knowledge — 60

4.5.4. ELF-control Knowledge	62
4.6. Architecture Overview	62
4.6.1. Input Stage	62
4.6.2. Selection Stage	64
4.6.3. Selection	64
4.6.4. Separation of Algorithm and Data Structure Selection	65
4.6.5. Output Code Generator Stage	66
4.7. Architecture Overview	67
4.8. Chapter Summary	68
References	**69**
5. The Input Stage	**71**
5.1. Input Stage Operation	71
5.2. Input Stage Rule Types	77
5.3. Chapter Summary	80
References	**81**
6. The Selection Stage	**83**
6.1. Selection Control Module	86
6.2. The Dependency Analysis Module	90
6.3. The Data Structure Designer Module	96
6.3.1. Representation of Data Structure Interdependency Information	97
6.3.2. Data Structure Representation During Selection	99
6.3.3. Data Structure Selection Operation	104
6.3.3.1. Router Domain Knowledge	107
6.3.3.2. Program Synthesis Knowledge	109
6.3.3.3. Design Interaction Knowledge	109
6.3.3.4. ELF-control Knowledge	111
6.3.3.5. Operations on the Interdependency Graph	112
6.4. The Algorithm Designer Module	115
6.4.1. Algorithm Representation	116
6.4.2. Algorithm Selection Operation	118
6.5. Chapter Summary	122
References	**125**
7. The Code Generator Stage	**127**
7.1. I/O Operation Synthesis	128
7.1.1. Input and Output Specification and Operation	129
7.1.2. Input Netlist Code Generation: An Example	129
7.2. The Use of Router Domain Knowledge in the Transformation Process	134
7.2.1. Effects of Applying Domain Knowledge	135

7.2.2. Domain Knowledge Driven Transformation: An Example	135
7.3. Stepwise Refinement in the Transformation Process	141
7.3.1. The Transformation Process	141
7.3.2. Transformation Comparison	145
7.4. Chapter Summary	150
References	**151**
8. Implementation	**153**
8.1. Implementation Characteristics	153
8.1.1. Input Stage	155
8.1.2. Selection Stage	155
8.1.3. Code Generator Stage	157
8.2. Design History	158
8.3. Modifying ELF: Is It Really Better?	158
8.3.1. How To Add a New Technology?	159
8.3.2. How To Add a New Algorithm Representation?	161
8.4. Issues in Debugging ELF-synthesized Code	164
8.5. Chapter Summary	165
References	**167**
9. ELF Validation	**169**
9.1. Experimental Methodology	170
9.2. Gate Array Style Routers	172
9.2.1. Comparison of ELF-synthesized Gate Array Routers	173
9.2.2. A Gate Array Routing Task	177
9.2.3. Comparison with Hand-crafted Code	178
9.3. Printed Circuit Board Style Router	179
9.3.1. ELF-Synthesized PCB Router	180
9.3.2. A PCB Routing Task	181
9.3.3. Comparison With a Production-Quality Router	184
9.4. Macro-Cell IC Style Router	185
9.4.1. ELF-Synthesized Macro-Cell IC Global Router	188
9.4.2. A Macro-Cell IC Routing Task	191
9.5. Chapter Summary	193
References	**195**
10. Conclusion	**197**
10.1. Summary	197
10.2. ELF: Hindsight and Evolution	199

References	**203**
Appendix I. Router Specification Manual	**205**
I.1. Syntax Description	205
I.2. Constraint Level Structure	206
I.2.1. Top-Level Constraint Specifications	206
I.2.2. Algorithm Constraints	207
I.2.2.1. Net_sorting	208
I.2.2.2. Node_sorting	208
I.2.2.3. Cost_function	209
I.2.2.4. Netlist	212
I.2.2.5. Output	214
I.2.2.6. Expansion	215
I.2.2.7. Net_composition	217
I.2.2.8. Routing_composition	217
I.2.3. Application Constraints	218
I.2.3.1. Type	218
I.2.3.2. Sub_type	219
I.2.3.3. Alg_type	219
I.2.3.4. Number_of_nets	219
I.2.3.5. Number_of_cells_per_net	220
I.2.4. Fabrication Constraints	220
I.2.4.1. Units	221
I.2.4.2. Pads	221
I.2.4.3. Connections	221
I.2.4.4. Xsize,Ysize,Zsize	222
I.2.4.5. Technology	222
I.2.4.6. PCB	222
I.2.4.7. Number_of_layers	222
I.2.4.8. Available_via_positioning	222
I.2.4.9. Layer	223
I.2.4.10. IC	224
I.3. Input constraint Schemes	226
Index	**231**

List of Figures

Figure 2-1:	A Router Taxonomy	15
Figure 2-2:	One-wire-at-a-time router	16
Figure 2-3:	Restricted-area router	17
Figure 2-4:	Maze Router Mechanics	20
Figure 2-5:	Expansion Example	22
Figure 4-1:	Router Generator Architecture	46
Figure 4-2:	ADL description: Interrelationship Segment	48
Figure 4-3:	ADL Algorithm Specification Example	49
Figure 4-4:	Data Structure Representation	51
Figure 4-5:	Multi-dimensional Data Structure Representation	52
Figure 4-6:	Use of Knowledge in Intermediate Representations	54
Figure 4-7:	ELF Synthesis Architecture	63
Figure 4-8:	Data Structure and Algorithm Designer Interrelationship	66
Figure 4-9:	ELF Architecture Knowledge Characterization	67
Figure 5-1:	User Specification Template (* required)	73
Figure 5-2:	Types of Domain Knowledge used in the Input Stage	74
Figure 5-3:	Sample Portion of a Specification Template	75
Figure 5-4:	Increased physical constraints imposed by diagonal wiring	76
Figure 6-1:	Selection Mechanism	84
Figure 6-2:	Types of Domain Knowledge in the Selection Stage	86
Figure 6-3:	Selection Control Module Process	87
Figure 6-4:	Knowledge in the Selection Control Module	88
Figure 6-5:	Knowledge in the Dependency Analysis Module	91
Figure 6-6:	Domain Knowledge in the Data Structure Designer Module	98
Figure 6-7:	Basic ELF Data Structure Interdependency Graph	100
Figure 6-8:	Data Structure Building: the Wavefront Data Structure	101
Figure 6-9:	Basic Data Structure Selection Possibility Tree	103
Figure 6-10:	Data Structure Selection Decisions	105
Figure 6-11:	Field Inversion	113
Figure 6-12:	Algorithm Designer Module Internal Algorithm Representation	117
Figure 6-13:	Algorithm Selection Decisions	118
Figure 6-14:	Selection Hazards: Premature selection	122
Figure 6-15:	Domain Knowledge in the Algorithm Designer Module	123
Figure 7-1:	Types of Domain Knowledge in the Code Generator Stage	128
Figure 7-2:	Sample BNF input file description	130

Figure 7-3:	BNF to C Code Result (Comments Manually Added)	133
Figure 7-4:	Effects of a Bounding Box on Search Area	136
Figure 7-5:	Code Generation Transformation Steps	143
Figure 7-6:	Transformation Example: ADL	146
Figure 7-7:	Example 1: User Input Specification	146
Figure 7-8:	Example 1: Sample Code	147
Figure 7-9:	Example 2: User Input Specification	148
Figure 7-10:	Example 2: Sample Code	149
Figure 8-1:	ELF Rule Breakdown By Stage	154
Figure 8-2:	Selection Stage Rule Breakdown by Module	156
Figure 8-3:	Code Generator Stage Rule Breakdown by Task	157
Figure 9-1:	Code Size and ELF Rule Firings for Gate Array Experiments	174
Figure 9-2:	Gate Array Experiment Decisions	174
Figure 9-3:	Gate Array Experimental Routers: Execution Statistics	177
Figure 9-4:	Synthesized Gate Array vs. Hand-crafted Router Comparison	179
Figure 9-5:	Different Via and Wire Traversal Cell Capacities	180
Figure 9-6:	Code Size and ELF Rule Firings for VAX Mainframe Experiment	182
Figure 9-7:	VAX Mainframe Board Experiment Decisions	182
Figure 9-8:	VAX Mainframe Board Experiment *grid* Data Structure	183
Figure 9-9:	DEC Easy Example Execution Statistics	184
Figure 9-10:	Comparison with a Production-Quality Router	185
Figure 9-11:	Experiment 5: DEC Easy Problem Global Routing Result	186
Figure 9-12:	Experiment 5: DEC Difficult Problem Global Routing Result	187
Figure 9-13:	Comparison Between Grid and Graph Routing Schemes	189
Figure 9-14:	ELF Rule Firings for Graph-Based Experiment	190
Figure 9-15:	Graph-Based Experiment Decisions	190
Figure 9-16:	Graph-Based Execution Statistics	191
Figure 9-17:	Effects of Blocking a Channel in a Graph-Based Router	192
Figure I-1:	Algorithm constraints: * required	207
Figure I-2:	Cost Function Determination	210
Figure I-3:	Input Netlist Definition Example	214
Figure I-4:	Output Routing List Example	215
Figure I-5:	Application constraints: * required	218
Figure I-6:	Fabrication Constraints	220

Figure I-7:	User Specification Template (* required)	225
Figure I-8:	Example 1: Global gate-array	227
Figure I-9:	Example 2: Detailed four layer PCB gate-array router	228

Preface

This book, and the research it describes, resulted from a simple observation we made sometime in 1986. Put simply, we noticed that many VLSI design tools looked "alike". That is, at least at the overall software architecture level, the algorithms and data structures required to solve problem X looked much like those required to solve problem X'. Unfortunately, this resemblance is often of little help in actually writing the software for problem X' given the software for problem X. In the VLSI CAD world, technology changes rapidly enough that design software must continually strive to keep up. And of course, VLSI design software, and engineering design software in general, is often exquisitely sensitive to some aspects of the domain (technology) in which it operates. Modest changes in functionality have an unfortunate tendency to require substantial (and time-consuming) internal software modifications.

Now, observing that large engineering software systems are technology-dependent is not particularly clever. However, we believe that our approach to dealing with this problem took an interesting new direction. We chose to investigate the extent to which *automatic programming* ideas cold be used to synthesize such software systems from high-level specifications. This book is one of the results of that effort.

This book describes a synthesis architecture for the automatic generation of technology-sensitive VLSI physical design tools. For this study we chose to focus on tools called *maze routers*, which determine that paths taken by wires in both integrated circuits and printed circuit boards. *Physical design* refers to the process of reducing a structural description of a piece of hardware down to the geometric layout of an integrated circuit. Successful physical design tools must cope with shifting technology and application environments. As technology progresses ever more swiftly, two bottlenecks form. The first bottleneck is in the management of the technological information: how we describe it, manipulate it, and so forth. The second bottleneck is the actual programming effort required to use the new technology: what algorithms and data structures we must pick to build robust, efficient software. In addition, the design software that we ultimately produce must be reasonably efficient. There are always inefficient, fragile, simple-minded schemes for adding new functionality to some design tool; we must strive to avoid such short cuts.

To attack these bottlenecks, we argue that the appropriate place for some of this technology-dependent information is not in the run-time environment of such tools, but in a *generator* for these tools. This book describes a synthesis architecture and its prototype implementation--called ELF--that integrates knowledge of the application domain with knowledge of generic programming mechanics. ELF strives to meet the demands of the target technology by automatically generating an implementation of the tool to match the application requirements. The ELF synthesis architecture has three key features. First, a very high level language, lacking data structure implementation specifications, is used to describe algorithm design styles. Second, application domain knowledge and generic program synthesis knowledge are used to guide search among candidate design styles for all necessary component algorithms, and to deduce compatible data structure implementations for these components. Third, code generation is used to transform the resulting abstract descriptions of selected algorithms and data structures into final, executable code. Code generation is an incremental, stepwise refinement process, and also relies on application domain knowledge, as well as generic program synthesis knowledge. ELF does *not* build software by selecting and glueing together fragments of executable code from some library. ELF synthesizes code *from scratch*, consulting only the input specifications for the tool, and a concise library of (non-executable) algorithm *styles*. The ELF prototype operates in the domain of VLSI wire-routing tasks. ELF demonstrates a synthesis architecture that efficiently generates router software using router knowledge and generic program synthesis knowledge as a synthesis guide.

We have walked a fine line throughout the writing of this book to keep it accessible to what we perceive to be two fairly different audiences. This book is aimed at researchers in the field of artificial intelligence, automatic programming, and software reusability. For these readers, we provide a wealth of detail about *how* ELF works, and describe enough of the domain in which ELF operates to impart some sense of what routers are all about, and how good ELF is at building them. The book should also be of interest to VLSI CAD tool designers and software system designers in general. For these readers, we have described our experience with how ELF was implemented, how successful it is in comparison to hand-crafted industrial tools, and how we might have attacked the design of ELF if we knew then what we know now. Concise background information is provided for both automatic programming and maze routing ideas. In short, we have done our best to insure that this breadth of readership is taken into account throughout this book.

We wish to thank a number of people who have provided invaluable assistance through the years. First and foremost are Dr. Elaine Kant of Schlumberger and Dr. Michel Doreau of Digital Equipment Corporation. Elaine provided support and guidance on artificial intelligence issues. Mike provided insight on real-world applications, including some wonderfully nasty-to-route benchmarks form Digital, and data about how Digital's production tools attacked these benchmarks. In addition, David Steier of Carnegie Mellon provided a sounding board for most of the ideas contained in this book. This research was supported by the CMU CAD Industrial Affiliates Program, which we gratefully acknowledge. Our synthesis tools and experiments were performed on equipment provided to Carnegie Mellon University under a grant from the Digital Equipment Corporation. We are also grateful for the use of facilities provided by the Semiconductor Research Corporation - Carnegie Mellon University Research Center for Excellence for Computer-Aided-Design, and the Department of Electrical Engineering, University of Pittsburgh, which made possible the production of this book.

And, of course, we wish to thank our spouses, Jay Strosnider and Martha Baron for suffering tremendous neglect during write and rewrite sessions.

Dorothy E. Setliff
Rob A. Rutenbar

**AUTOMATIC PROGRAMMING
APPLIED TO VLSI CAD SOFTWARE:**
A Case Study

Chapter 1
Introduction

One of the basic problems in technology-dependent software tools is the placement, integration and maintenance of technology-specific code within the tool. Accommodating new technologies, or versions of old technologies, places a burden on the design and implementation of technology-sensitive tools. Technology shifts so quickly that the development of new tools, or re-engineering of old tools, often lags behind the introduction of new technologies. A method of reducing the cost of this burden is to produce technology-independent tools. Successful design tools always achieve some measure of technology independence, permitting them to address an acceptably wide range of design tasks, and allowing at least minor changes in problem specifications, such as geometric, electrical, or fabrication constraints. However, it is safe to assert that technology independence--when it is achieved at all--is still accomplished largely through ad hoc techniques, and is still of immense practical interest.

Informally, we can identify two extremes among the conventional solutions proposed to handle the general technology-independence problem. One extreme is to construct a "general-purpose" tool. By "general-purpose" we mean simply that the tool is designed with the ability to cope with a variety of technologies as well as a variety of problem specifications. The methods used to handle the variety of cases might vary, e.g., strict parameterization, expandable data structures and algorithms.

A difficulty with this approach is the problem of introducing a new variant. Handling these additions usually requires massive changes such as new data structures to handle efficiently the change in requirements, or substantial algorithm modification. Implementation techniques to adapt the tool to the problem range from simple schemes, such as embedded conditionals or conditional compilation, to complex schemes, such as threaded-code segments at each technology-dependent decision point in the program code with selection occurring at run-time, compile-time, or both. However, none of these approaches provides an elegant solution to the problem of introducing a new variant. Handling these additions typically requires substantial changes which must be designed, coded, integrated and tested, to meet stringent performance demands. As new variants are introduced, such tools grow by accretion over their lifetime, evolving into large baroque systems, which are difficult to maintain, update, modify and extend with changing technologies. The performance of such tools slowly degrades as the system grows. Moreover, the code resulting from such approaches may still fail to meet some critical nuances of the target technology [2], resulting in a second round of *ad hoc*, hand-crafted tuning of large complex software.

The other extreme solution to the technology-independence problem is to construct a special-purpose custom tool for each different technology or problem specification. In effect, the tool designer abandons the idea of a general tool that adapts to each problem, and instead writes a new program for each new problem. This approach produces a tool that is highly tuned for a specific problem and technology. An example of such a tool is a VLSI router developed internally by a silicon foundry for use with its own proprietary technology, e.g., a router for a fixed family of semi-custom chips, such as gate arrays [1].

These two approaches are radically different. The special-purpose tool offers greater performance than does the general-purpose tool, but at a much higher development cost. Advances in technology could invalidate the tool requiring an entirely new special-purpose tool to fit a new problem constraint. A general-purpose tool can also be extremely difficult to expand: not only must problems related to new functionality be solved, but they must also be worked cleanly into the prior tool implementation. Moreover, it can also be argued that general-purpose tools, because they are not highly-tuned for any specific task, trade away some performance in favor of flexibility.

These conventional approaches fail to address the problem of technological evolution and its effect on tool *design*. This book explores the novel technique of automatically generating a technology-dependent tool based upon a technology description. This technique places the burden of the technology-dependent design problem in the **generation** of a tool, rather than in the execution of a tool. A special-purpose implementation of the tool is automatically generated to match the problem to be solved. This technique develops software technology by which design algorithms can be transformed into flexible, retargetable tools.

A generator for custom design tools offers several advantages. For each new design task, a tool is generated automatically to closely mirror the problem specification, and as such does not carry the run-time baggage of a general-purpose tool (a generated tool need only carry that information required to solve its targeted problem, and nothing else). We contend that the reduction of run-time baggage will result in performance more comparable to a special-purpose custom design tool, but without the large development costs. In addition, problems related to specific technology changes need to be addressed only once, in the generator for the tool. Once the changes are implemented in the generator, the solution can be invoked, instead of reinvented and reimplemented, for each new tool implementation that must deal with this technology.

1.1. The Application Domain

The impact of technological evolution is pronounced in physical design tools for integrated circuits. The development of new integrated circuit fabrication technologies often outpaces the development of VLSI physical design tools capable of utilizing the benefits of the new fabrication technology.

As a simple example of the problem of tracking a changing technology environment, consider the software for a detailed router for a semi-custom integrated circuit technology, e.g., a gate array or sea of gates chip. Changing a design rule related to fabrication, e.g., decreasing the minimum metal routing layer width from 2 microns to 1.8 microns, should not be a problem for a router that parameterizes metal-width internally. But if this same router was custom-designed to support a one-poly-one-metal process, it may not be able to support the addition of another metal layer or, worse, two more metal layers. If this router was not designed to accommodate the addition of new non-interacting wiring layers, extending it to do so may be very difficult. Similarly, if the router was designed strictly to handle rectilinear wiring restrictions, then extending it to handle diagonal wiring may be very difficult. The reason is that it no longer suffices to change the value of some pre-determined router attribute; rather, we may need to change the router's basic architecture, algorithms and data structures.

ELF, a system for transforming flexible, high-level specifications for integrated circuit (IC) and printed circuit board (PCB) routers into working router software, demonstrates the feasibility of the generator approach. ELF provides a new automatic programming environment that integrates expertise about the mechanics of building good routers with automatic program synthesis techniques. This synthesis architecture efficiently writes router software using domain-specific router knowledge and generic program synthesis knowledge as a synthesis guide.

ELF generates working routing software for a wide variety of technologies, from multiple-layer PCBs to semi-custom ICs. ELF demonstrates that real, workable CAD tools, in particular routers, can be synthesized automatically from high-level specifications. These results support the argument that the integration of specific knowledge of the target domain within a program synthesis environment is essential for synthesizing real-world design tools.

We do not claim that the ELF synthesis architecture is the panacea to software development woes. Rather we claim that for some types of software, in particular technology-dependent software, such an approach is valid. Several characteristics are required of the software domain to apply successfully the

ELF synthesis architecture. Specifically, these characteristics are:
- **Stability:** the ideas, algorithms, data structures, optimizations, etc., underlying the application have matured, and are widely understood.
- **Sensitivity:** minor changes in the target technology mandate substantial changes to the code itself, either the coarse structure (e.g., basic algorithm choices) or the fine structure (e.g., element layout of data structures, or code optimizations).
- **Configurability:** by changing the coarse and fine structure of the code, one can retarget the tool to numerous, widely varying applications; that is, the fundamental CAD techniques themselves are *practical* and *useful* across a large set of problems, but the software is difficult in practice to re-engineer for each new problem.

These requirements guarantee that the emphasis is on tool generation rather than on new algorithm formation. These sorts of design problems are often referred to as *routine design* task, precisely because they require re-engineering of known software algorithmic approaches, rather than inventing radically new solutions. However, it is critical to note that routine designs are not necessarily easy or trivial designs. The fact that the next 10,000 lines of code looks much like the last 10,000 lines--except that all the algorithms and data structures are changed--clearly does not mean this new code can be produces in an hour.

1.2. Knowledge Sources

Many of the ideas for program generators originate, at least in part, from the field of automatic program synthesis. The program synthesis field has itself been expanding, though to date the synthesis of large, real-world application programs like CAD tools has not occurred. Chapter 3 reviews the relevant approaches within the program synthesis field. We characterize the current approaches to program generation according to their primary knowledge source.

While program generators often utilize several sources of knowledge representing a continuum of approaches, two primary knowledge sources exist: domain-independent programming knowledge and domain-related knowledge.

The first knowledge source, domain-independent programming knowledge, is knowledge about implementation code syntax and data structures. Syntactic knowledge focuses on performance characteristics and implementation mechanisms for data structures and algorithms. But the peculiarities of a particular application domain are not incorporated in this approach. As an example of knowledge not incorporated in this approach, suppose that in a particular VLSI routing task the linked-list form of a required data structure is always known to be preferable. This knowledge is simple expertise, possibly acquired as the result of many previous attempts to implement this data structure in other, less optimal, styles. The lack of application-specific knowledge weakens program synthesis when applied to real-world problems. Still, program synthesis research based on domain-independent programming knowledge has demonstrated, for example, the viability of choosing data structures using data structure specific knowledge as well as the usefulness of *generic* syntactic knowledge.

The second knowledge source, domain-related knowledge, is knowledge about the application domain. Generally, the application is severely restricted within the program generator. Program synthesis research based on application domain knowledge supports the viability, at least in a fairly narrow application domain, of using domain knowledge in program synthesis.

We integrate the key features of both approaches--problem-specific domain knowledge and program synthesis techniques--into one program generator targeted for a real-world VLSI design application. In particular, we utilize problem-specific domain knowledge and data structure implementation knowledge as two sources of program synthesis knowledge. These two knowledge sources are used to help select the right approach to implementing the desired design tool, and to help select data structures, algorithms, optimizations, and domain-dependent code "tricks." Generic program-synthesis syntactic knowledge is then used to generate the actual target code implementation. We demonstrate that program synthesis is viable in real engineering applications, where incorporation of detailed knowledge of the domain is required for successful, efficient implementations.

1.3. Book Organization

The remainder of this book is organized as follows. Chapter 2 reviews maze-routing algorithms, the chosen domain for our work, and presents justification for this choice. Chapter 3 presents a survey of relevant results from prior work on program synthesis, placing our work in the context of the larger field of software reusability. Chapter 4 gives an architectural overview of our router program generator, including a high-level description of the three main modules, and introduces our current implementation, called ELF. Chapter 5 discusses the first of the three modules, the Input Stage and related user interface knowledge representation issues. Chapter 6 discusses the second of the three modules, the Selection Stage and its related data structure and algorithm selection as well as representation issues. Chapter 7 discusses the final stage, the Code Generator and the domain knowledge-driven transformation process. Chapter 8 details the implementation characteristics of the ELF system including domain knowledge placement and the structure of ELF's rule-based implementation. Chapter 9 validates the ELF architecture using illustrative ELF-synthesized router examples ranging from gate arrays to multi-layer PCBs. Finally, Chapter 10 reviews the contributions of this work and also discusses some future directions for work in automatic programming.

References

[1] D. Hightower, "The Lee Router Revisited", *ICCAD-83*, IEEE, 1983.

[2] R. Newton, "A D&T Roundtable: CAD Software Development", *IEEE Design and Test of Computers*, August 1988.

Chapter 2
Application Domain: Routing Algorithms

Our work on program synthesis is targeted at a particular VLSI CAD problem: wire routing. We focus on a particular class of routing tools called *maze routers*. This chapter describes the actions of routers in general and maze routers in particular, and then reviews the relevant mechanics of maze routing systems. Maze routing algorithms have been applied to a variety of wire routing problems. Varying "environmental" constraints strongly affect a router's implementation. As evidence of this dependency on environmental constraints, we organize the constraints into three fields: algorithm, application, and fabrication constraints. We then describe the effects of modifying various constraints within these fields.

2.1. Routing Algorithms

A router determines the placement of the wiring paths that connect electrical components, for example: chips on a printed circuit board, or gates on an individual chip. Pins or terminals on any electrical components that are to be wired together are part of a set called a *net*. Nets are wired within the environmental constraints of the chosen application. Such constraints include the number of physical wiring layers, whether layers can cross without interaction, the actual physical dimensions of the electrical connections within the component, and the placement of physical connections between non-interacting wiring layers, called *vias*. In addition, there are performance related requirements common to many routing tasks, such as finding the shortest possible connection to maximize the speed of the circuit, or to minimize noise. A router must also consider constraints such as the minimum spacing between wires and the placement of wires routed earlier, which may block subsequent attempts to wire a given net. These constraints often are contradictory and result in a routing problem which is well explored [2], [4], [6], [7], [8], [9], [12], [13], [14], [16], [20], [21], but can be quite difficult. A successful router will route all nets without violating any physical technology constraint, or any overall performance-related constraint.

2.1.1. Router Classifications

Maze routers form one broad algorithmic approach to routing tools. Other different algorithmic approaches also exist because of the existence of significantly different routing tasks, and because particular routing approachs offer distinct advantages and disadvantages. Figure 2-1 shows a simplified router taxonomy derived from [16].

We discuss four common varieties of routers: one-wire-at-a-time routers, restricted-area routers, global routers, and detailed routers.

One-wire-at-at-time routers

These routers route wires in a sequential fashion one-at-a-time, with no restriction on the shape or occupancy of the routing area. A wire ordering function chooses the next wire to be routed. The wiring ordering function varies

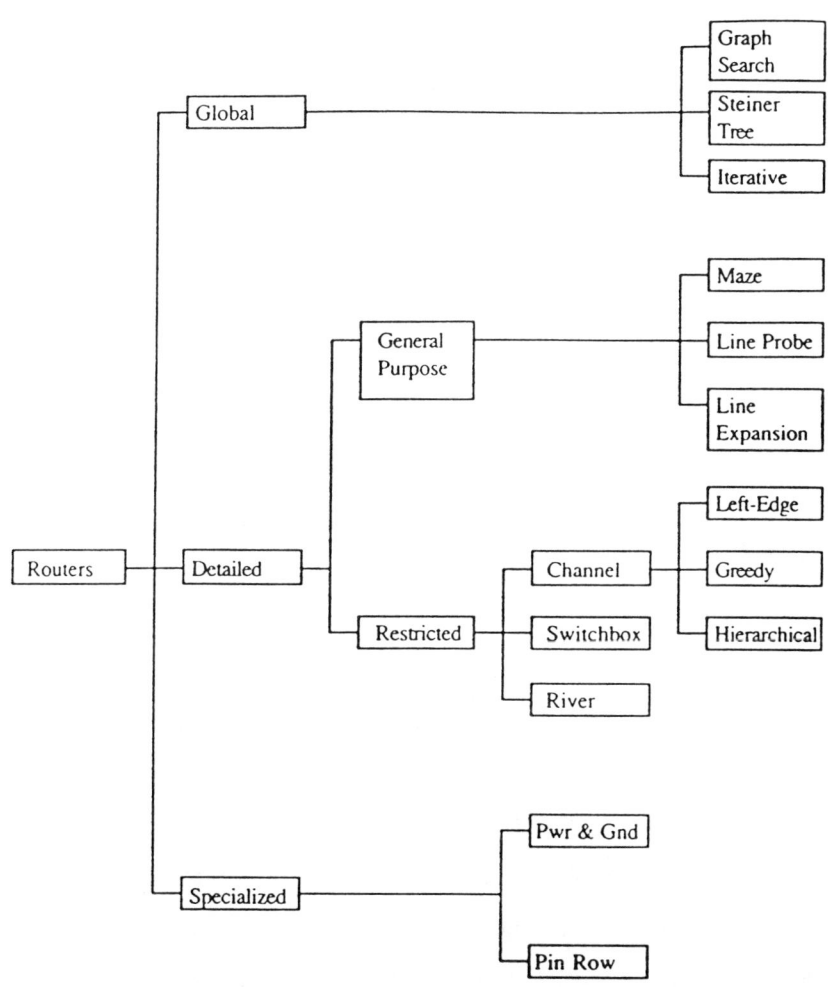

Figure 2-1: A Router Taxonomy

from implementation to implementation to reflect different problem requirements. Maze routers [11, 14, 18, 20] and line-probe routers [8] are examples. An example is shown in Figure 2-2. Objects 1-2 are connected via the wire labeled A, and Objects 3-4 are connected via the wire labeled B. A one-wire-at-a time router selects a single wire (in this case, the one labeled A) and routes it. The wire labeled B is then routed.

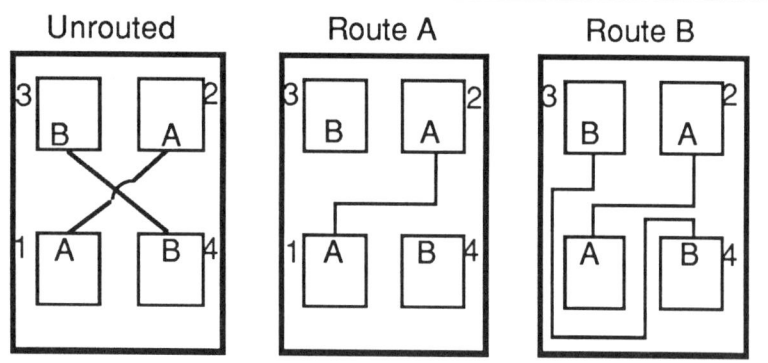

Figure 2-2: One-wire-at-a-time router

Restricted-area routers

In some IC applications the overall routing task is usually decomposed into many small routing tasks. For example, an IC chip is often decomposed into routing regions, with one routing task for each routing region on the chip. In these cases, net interaction is the main problem. One-wire-at-a-time routers, such as maze routers, do not satisfactorily solve these restricted-area routing problems. Restricted-area routers specialize in one specific routing area geometry. Channel and switchbox routers are examples [2, 10, 17]. Rectangular routing regions in which nets enter through only two opposite sides of the region are called *channels*; if nets can enter from all four sides, we refer to the region as a *switchbox*. The common characteristic of these routers is the simultaneous consideration of all nets and their wiring interactions as routing progresses. The resulting routing is essentially net-order independent. An example of this type of router is shown in Figure 2-3. This is a channel router. This router simultaneously looks at the relative placement of terminals (labeled a-d in Figure 2-3) for all the nets in deciding the interconnect route.

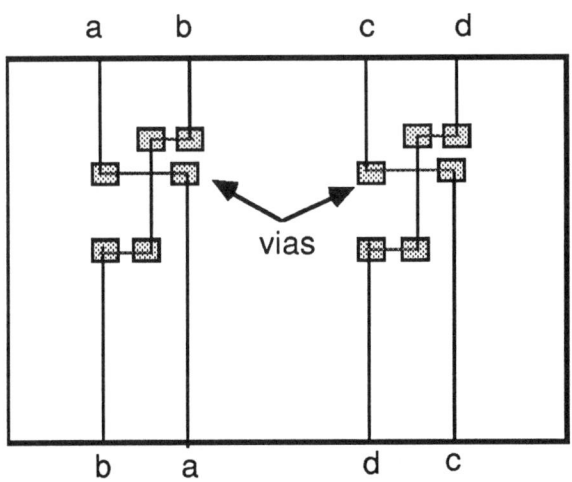

Figure 2-3: Restricted-area router

Global routers:

A routing problem is often organized hierarchically [1, 20]. First, the total routing area is divided into regions. Second, the nets are routed through the regions, but actual placement of the nets onto exact physical locations within each region is not performed. Third, because we now know which nets must cross each region, each region is then separately routed to derive the actual physical placement for each net. The routing of nets through regions is performed by the global router. The main consideration is the minimization of congestion bottlenecks, i.e., we try to avoid having too many wires trying to cross a region, which makes the final, detailed routing of the region impossible. Global routers tend to be one-wire-at-a-time routers. The algorithm may be either a line-probe [8] or maze router type or both.

Detailed routers

A detailed router assigns the exact physical placement of each wire segment and via in each net [17, 9]. Region routers, such as channels or switchboxes, are always detailed routers. However, one-net-at-a-time routers, especially maze routers, are often used both for global routing and detailed routing. In addition, detailed maze routers are often used as a last resort router to clean up after other routers fail.

2.1.2. Types of Routing Constraints

Three broad types of environmental constraints influence the design of a specific router. These constraints pertain to fabrication, application, and algorithm selection. Fabrication constraints pertain to the technology of the object, often referred to abstractly as the *carrier*, to be routed. An example of a fabrication constraint is the number of routing layers available in the carrier. Application constraints pertain to the tool user's differing requirements for exactly what job, or jobs, the tool is to perform. An example of an application requirement is that the router will be used for global routing of two-layer gate arrays, or detailed routing of four-layer PC boards. Algorithm selection constraints pertain to the overall strategy employed in a particular router; these constraints reflect the user's preferences or expertise about which particular variant of a routing algorithm is likely to work best for a specific application with specific fabrication constraints. Examples of algorithm selection are the longest-net-first ordering schemes for a one-net-at-a-time router, and the penalty functions used to measure wire congestion for a global router.

2.2. Application Domain: Maze Routers

Maze routers form a broad and adaptable class of routers also known as Lee routers, or Lee-Moore routers, or flood routers [11, 14, 20]. Maze routers can be implemented for both global and detailed routing tasks. For example, global routers for both custom and semi-custom ICs, as well as large PC boards, are often based on maze-routing algorithms. Detailed routers for semi-custom applications like gate arrays, and PC boards, are also common. Before we present arguments in Section 2.4 about why maze-routers are a suitable target for tool generation research, we first describe the basic mechanics of a simple maze router.

Consider the following routing problem: route a set of nets each consisting of a set of terminals, one at a time, in some arbitrary order. A maze router is used when the shortest connection path or, more generally, the least-cost connection path is desired. Because only one wire is routed at a time, and previously routed wires become obstacles for future wires, the order of routing these nets is crucial. The simplest possible routing problem here is to connect nets that are constrained to have only two terminals (two-point nets) in a single conducting layer (hence, no vias), while attempting to find the shortest possible routing path. There are four phases in this simple maze router: setup, expansion, backtrace, and cleanup. These phases are illustrated in Figure 2-4 and discussed in turn, below.

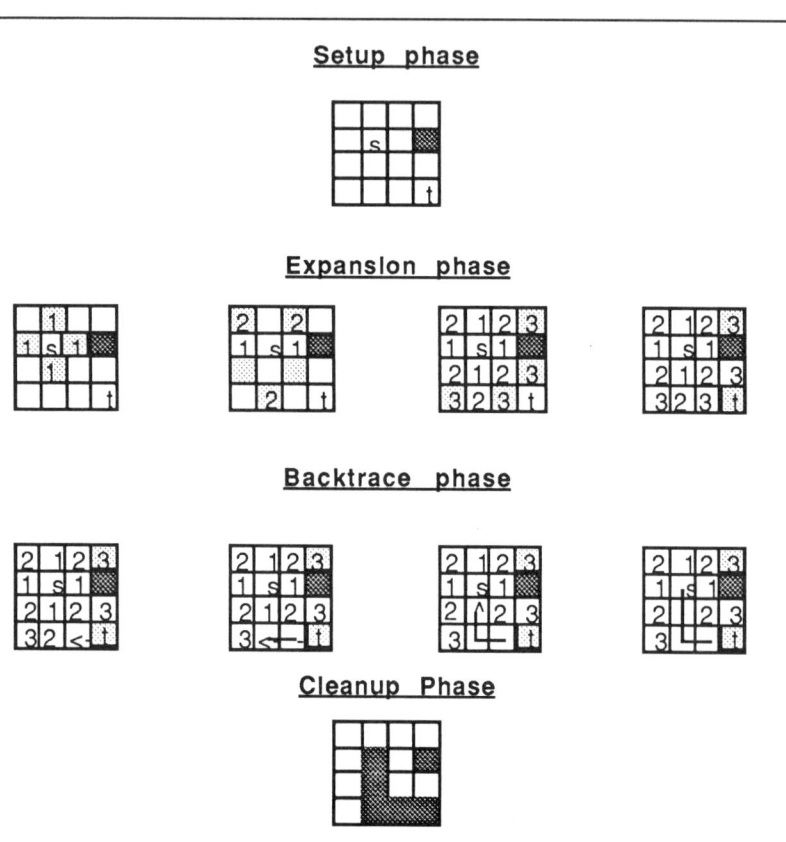

Figure 2-4: Maze Router Mechanics

In the *setup* phase, shown in the first row of Figure 2-4, the routing area is forced onto a grid. Each net is composed of a path through a sequence of cells on this grid. For our restricted example, each location on the grid represents an area through which just one wire can pass. In more complex examples, each location of the grid may represent an area through which many wires and vias, which allow connections between layers, may pass. Each grid location can either be empty, occupied by a wire or by a placed component, such as a chip or gate or cell. If occupied by a wire, or a placed component, the grid location is blocked, and is not available for finding a wiring path. In addition, individual grid locations, usually referred to as *cells*, can have their own costs associated with them. This cost measures the incremental penalty that accrues to a wire path that tries to cross this cell. In general, cell cost mechanisms can be very complex. But for this simple example, we assume unit-cost, i.e., the cost of traversing any cell is 1, and hence, the final cost of a path is the *number* of cells it contains, which is precisely its length. Hence, a unit-cost router attempts to find the shortest routing path.

The next phase, shown in the second row in Figure 2-4 is the *expansion* phase. The basic idea is to start from one terminal of the net to be wired, the *source* cell, and search outward until we find the *target* cell, representing the other end of the net. In our example, there are only two-point nets, so choosing which cell is the source and which is the target is trivial. We implement this search by examining all paths a distance of one cell from the source, then a distance two cells from the source, then three cells, and so forth. At any time in this search, the cells being examined constitute a *wavefront* of cells that expand outward from the source cell. Because, in our example, we have assumed unit cost cells, expansion becomes equivalent to evaluating the distance traveled. We borrow some conventional search terminology to describe the process [15]. For example, when we use a cell on a wavefront to determine the shortest path to reach its neighbors, we say the cell has been *expanded*.

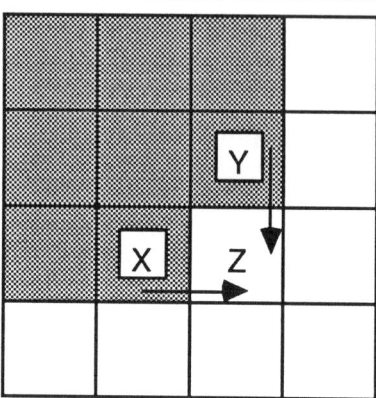

Figure 2-5: Expansion Example

The basic computation that occurs during expansion is illustrated in Figure 2-5. A particular cell z is adjacent to cells x and y that have already been *reached* (i.e., we know the minimum cost path from source to x and to y), cells x and y are already on the wavefront. We now wish to reach z, to determine the minimum cost to z. This can be computed easily; the minimum cost to reach z is just the minimum cost to reach x plus the cost of traversing z itself. The essential points are that, if we have good data structures to store the cells on the wavefront, and good algorithms to find the next wavefront cell to expand (in this case, x) and the next adjacent cell (in this example, z) to reach, the expansion process requires only knowledge local to these cells: the cost of traversal for z, and the known costs to reach its expanded neighbor x.

More generally, there are several options for performing the expansion process [20]. Again, conventional terminology suffices to define these options.

If we search all cells at distance i from the source cell before looking at cells at distance $i + 1$, we are performing a *breadth-first* search. If, instead, we search preferentially those cells that we know are "closer" to the target (e.g., by evaluating an estimated distance from this cell to the target), we perform an ordered *depth-first* search. In general, we expand the least costly cell on the wavefront, a *best-first* search, assuming that individual cells in the grid have their own individual cost to traverse them. In this expansion search, the algorithm is quite similar to Dijkstra's algorithm [3, 22] for finding the least costly path between two nodes in a weighted graph.

After the target cell has been reached, we know there is a source-to-target path in the grid. We can now proceed with the *backtrace phase*, shown in the third row in Figure 2-4, to find the actual wiring path. In the simplest case, as each cell is reached, information is embedded in the cell indicating the direction of expansion used to enter the cell. In the previous example, this would mean remembering that the shortest path to z went through x, and embedding, in effect, a pointer in the cell to the predecessor cell on the least cost path during the expansion phase. This information is used to define the wiring path during the backtrace phase. In particular, this directional information can be traced backwards from the target cell, through each cell added to the wire list, until the initial source net cell is reached; in this way we trace out the final wiring path backwards, from target to source.

Finally, the *cleanup* phase, shown in the bottom row in Figure 2-4, reinitializes the grid to show the existence of this new wire connection as a blockage on the relevant wiring layers. In particular, we must remove any information embedded in the grid that would interfere with the expansion phase of subsequent nets. After cleanup, the routing grid is then ready for a new net; the grid has only routable and blocked cells labeled. These four phases are then repeated for each remaining net to be routed.

2.3. Maze Router Varieties

The simple maze router described in the previous section is only one variant in the space of possible maze router implementations. Its basic constraints, e.g., single conducting layer, two-point nets, unit-cost breadth-first expansion, can all be relaxed. However, just as in the previous discussion of constraints on general routing schemes, a given maze router implementation must address constraints arising from three sources: fabrication, application, and algorithms.

This section briefly surveys the effects of technology constraints on maze routing implementation techniques. We attribute the differing characteristics of maze router implementations to the fabrication, application and algorithm constraints from which they arise. However, it is important to note that not all constraints are fixed and inflexible, and that similar effects on router implementation can be attributed to very different environmental constraints. For example, a restriction to two-point nets might arise due to real technology limitations (e.g., in high-speed mainframes, nets are often constrained like this to manage more carefully their impact on signal delay). On the other hand, the technology may support multi-point nets, but the user who configures the router for the application may *prefer* to decompose into two-point nets, based on some expertise about the likely consequences of this choice.

2.3.1. Fabrication Constraints

Fabrication constraints pertain to the physical environment of the carrier which the router operates. Some fabrication constraints that result in different maze router implementations are:

Multiple routing layers

The addition of multiple routing layers complicates routing. In particular, the capability of changing layers, and exploring paths with segments on multiple layers, complicates the expansion phase. Support for multiple layers make take the form of simply providing support for two layers, forcing carriers with many more layers to be routed by partitioning the set of input nets among layer *pairs*, and then routing each layer pair separately. On the other hand, explicit support for more than two layers can be built in to the expansion phase.

Preferred routing direction per layer

Each layer may have a preferred routing direction that is determined by the fabrication process. (Alternatively, the directionality can be a user preference, instead of a technology constraint. Expansion in this direction is strongly suggested, but violations to complete nets are allowed.)

Vias

With support of multiple layers must come support for vias, which allow connections among layers. Vias may have fixed or floating placement restrictions on the carriers, and some regions may be barred from having any vias at all. In addition, since in most technologies vias are larger in diameter than wires, via *exclusion* rules must be enforced: it may not be possible to put vias in some adjacent cell configurations on the routing grid. With very many layers, via *stacking* rules must be enforced: it may not be possible to put a via between layers 1 and 2, and a via between layers 7 and 9, vertically above each other in the same grid location.

Wiring angle selection

The definition of adjacency can be expanded from Manhattan geometry. For example, we can include 45 degree cell adjacency to find paths with 45 degree bends. This is only valid under the assumption that the physical implementation allows such non-Manhattan connections. The definition of adjacency greatly affects the internal representation of the routing problem.

2.3.2. Application Constraints

Application constraints pertain to router operation within any given application. Some application constraints that result in different maze router varieties are:

Router classification

A maze router may be either a global or detailed router. Obviously, this has a substantial impact on the router's data structure and algorithm implementations. For example, at one extreme, a detailed router might require a very large detailed grid for expansion, while a global router may not operate on a grid at all, but on a graph representing adjacent routing regions (as in a custom macro-cell routing task).

Net definition

A maze router may route either two-point net connections alone, or a set of multi-point nets. Selection of the first may require algorithms to make the best net decomposition into two-point nets, if in fact the input nets are not already presented in the form of two-point nets.

2.3.3. Algorithm Constraints

Algorithm constraints affect router efficiency and also routing quality. These choices may result from a change in technology, or from new algorithm advances. Other choices may arise as a result of the desired algorithm performance tradeoffs, such as speed versus space. Some algorithm choices that affect router efficiency are:

Net ordering

The order of nets chosen to be routed is important. In some applications, it might be advantageous to route the longer nets first (e.g., timing critical length nets), while in other applications, the shorter nets (e.g., critical length nets) might be considered to have top priority. Sorting algorithms may need to be included.

Expansion path definition

The maze router may be tuned to find either the shortest path, or the path with the least cost. Note that the shortest path may not be the path with the least cost.

Expansion search selection

Maze routers can use breadth-first, depth-first, or best-first expansion algorithms, all with differing results.

Cell Reentrancy

The expansion search phase may be tuned to allow multiple entrancies into a single cell at different times. This has the advantage of simplifying the cost function as it need only be concerned with the cost of entering a cell, not with cost of traversing a cell. It has the disadvantage of reexploring old, non-useful paths.

Wavefront data structures

The ability to find quickly the next cell to expand, remove old expanded cells, and find the cost of reaching a new cell, rests largely on the choice of the proper wavefront data structure to match the requirements of the expansion search.

Detour limits

The amount by which a routed net exceeds its minimum cost or length is often referred to as its *detour*. Some routers limit this detour to limit the spatial extent of a net, using the following well-established rule: in a good placement with sufficient wiring space, most nets are routed near their minimum cost or length. Limiting the detour can be done, for example, by placing a frame of heuristically chosen size around the terminals of the net, and expanding no cells outside the frame. This has the effect of reducing the search space and accelerating the expansion phase. We may actually settle for a slightly longer or more costly net, but one that does not meander widely around the carrier, possibly congesting other nets to be routed.

Backtrace optimization

It has been argued [8] that path optimizations applied during backtrace can greatly improve wiring quality, for example, by removing useless kinks and bends that serve mainly to reduce via availability for future wires. Because expansion typically takes orders of magnitude more time than backtrace, this argument suggests optimizing the expansion for speed, and applying path modifications later during backtrace.

2.4. Why Choose Maze Routers?

Our overall goal is to investigate the feasibility of applying program synthesis methods to the design of technology-sensitive CAD tools. The domain of choice is maze routing tools. There are several characteristics that the candidate application domain must meet to accomplish successfully this research. The first requirement is that there must be a variety of technology-dependent applications. In particular, these applications must be concisely and unambiguously specifiable. The previous survey of maze router applications, in particular, the effect of what we termed fabrication and application constraints, demonstrate the wide range of maze router variants.

The second requirement is the desirability for stable, mature domain algorithms. This ensures the research focus is on tool generation rather than on new algorithm formation. Maze routers rely on a small kernel of basic algorithms, e.g., for different expansion schemes, different backtrace schemes, etc. These algorithms are, by themselves, fairly simple, but they have been elaborated and adapted over the last few decades [19, 11, 20, 18, 8, 2, 5, 12, 13, 14] to attack new routing problems.

The third requirement for this work is that complex elaborations of the simple algorithms are needed to adapt them to specific applications. This requirement ensures, in effect, the non-triviality of this approach: the domain must be too complex to permit glueing together simple algorithm fragments to produce good routers. Taken together, we regard these characteristics of the class of maze routing algorithms as being extremely well-suited to our requirements for tool synthesis research.

2.5. Chapter Summary

This chapter presented a router taxonomy and described in detail the algorithmic behavior of maze routers. Maze routers are used in a variety of applications. These applications present differing sets of environmental constraints. We organized these environmental constraints into three fields--algorithm constraints, application constraints, and fabrication constraints--and we reviewed the effect of modifying each of these constraints on the maze router implementation. We presented all this detail to justify the choice of maze routers as a good target for work on program synthesis: maze routers are mature, widely-used, technology-sensitive tools able to be configured to attack an impressively wide array of wire routing tasks. Before describing our strategy for building a maze router software synthesis tool, we review in the next chapter a variety of software synthesis ideas, and then revisit router synthesis in Chapter 4.

References

[1] M. Burstein and R. Pelavin, "Hierarchical Wire Routing", *IEEE Transactions on Computer-Aided Design*, Vol. CAD-2, October 1983.

[2] D. Deutsch, "A 'Dogleg' Channel Router", *Proceedings 13-th Design Automation Conference*, IEEE, 1976.

[3] E.W. Dijkstra, "A Note of Two Problems in Connexion with Graphs", *Numer. Math.* 1959, pp. 269-271.

[4] J. Dion, "Fast Printed Circuit Board Routing", *24nd Design Automation Conference*, 1987, pp. 727-734.

[5] M. Doreau, "Private Correspondence, February 1989".

[6] D.A. Edwards, "The Architecture of the Manchester Routing Engines", Tech. report, Univeristy of Manchester, England, September 1988.

[7] A.C. Finch, K.J. Mackenzie, G.J. Balson, and G. Symonds, "A Method for Gridless Routing of Printed Circuit Boards", *22nd Design Automation Conference*, 1985, pp. 509-515.

[8] D. Hightower, "The Lee Router Revisited", *ICCAD-83*, IEEE, 1983.

[9] J. Hoel, "Some Variations of Lee's Algorithm", *IEEE Transactions on Computers*, Vol. c-25, January 1976.

[10] R. Joobani, *Weaver*, PhD dissertation, Carnegie Mellon University, 1985.

[11] C. Lee, "An Algorithm for Path Connections and Its Applications", *IRE Trans. Electronic Computers*, September 1961, pp. 246-265.

[12] R. Linsker, "An Iterative-Improvement Penalty-Function-Driven Wire Routing System", *IBM J. Res. Develop*, Vol. 28, September 1984.

[13] R. Nair, et. al, "Global Wiring on a Wire Routing Machine", *19th Design Automation Conference*, IEEE, 1982.

[14] T. Ohtsuki, ed., *Layout Design and Verification*, Elsevier Science Publishing Company, Advances in CAD for VLSI, Vol. 4, 1986, ch. 3.,

"Maze-Running and Line-Search Algorithms".

[15] J. Pearl, *Heuristics Intelligent Search Strategies for Computer Problem Solving*, Addison-Wesley Publishing Company, Inc., 1984.

[16] B. Preas and M. Lorenzetti, *Physical Design Automation of VLSI Systems*, The Benjamin/Cummings Publishing Company, Inc, 1988.

[17] R.L. Rivest and C.M. Fiduccia, " A 'Greedy' Channel Router", *19th Design Automation Conference*, IEEE, 1982.

[18] F. Rubin, "The Lee Path Connection Algorithm", *IEEE Transactions on Computers*, Vol. c-25, September 1974.

[19] S. Rubin, *Computer Aids for VLSI Design*, Addison-Wesley, 1987.

[20] J. Soukup, "Circuit Layout", *Proceedings of the IEEE*, Vol. 69, October 1981.

[21] H. Terai, M. Hayase, T. Kozawa, "A Routing Procedure for Mixed Array of Custom Macros and Standard Cells", *22nd Design Automation Conference*, 1985, pp. 523-528.

[22] P.D. Whiting and J.A. Hillier, "A Method for Finding the Shortest Route Through a Road Network", *Operational Research Quarterly* 1959, pp. 37-40.

Chapter 3
Software Reusability

This chapter locates our work within the broader context of research in automatic programming and, more generally, software reusability. Work on automatic programming strives to generate code from abstract, though not necessarily executable, descriptions. Work on software reusability, which can be regarded as encompassing automatic programming, strives to devise methods, architectures and tools to allow knowledge about one software system to be *reused* to build a different software system. A complete survey of ideas and techniques here is beyond the scope of this book; see Rich [12] for a survey of current work in automatic programming in general, and Biggerstaff and Perlis [3] for a comprehensive and recent treatment of the software reusability field. For our purposes, we can use the basic taxonomy of the field presented in [3] to clarify the methods and goals of our own work on ELF. The two broad categories of work on reusability are *composition-based systems* and *generation-based systems*.

3.1. Composition-Based Systems

Composition-based systems strive to build software systems by composing resuable components, e.g., software modules, libraries of parameterized code, abstract templates for code, etc. When applications are known to be composed primarily of "common" parts, such approaches can provide considerable leverage, because the available building blocks remove the need to write much low-level code. However, not all applications have this network-of-black-boxes character. In the CAD world, the most obvious and visible manifestation of these ideas is the object-oriented programming paradigm [4, 11]. Here, data objects and the procedures that manipulate them are packaged as units and made available to system designers. New objects can be created by composing old objects, and there are even sophisticated provisions for these child objects to inherit properties of their parents.

However, as has been noted of application software in general [3], and CAD software in particular [10], object-oriented techniques are a useful tool, but not a complete solution. Complex applications usually involve a large spectrum of subtle computations and ingeniously-contrived data structures. Although a library of reusable objects may provide numerous low-level pieces for such an application, it is unlikely to provide these highly problem-specific pieces. Further, an attempt to create a fully abstracted, reusable object for each component in the solution is likely to be a time-consuming, but not well-leveraged activity: the results may be so specific that they are not *useful* in another application, even a re-engineered version of the application they were originally built for. Worse, there is often a tension between nicely abstracted software components, with many layers of hierarchy and information-hiding, and components that must meet tight performance criteria for speed or space. Some forms of composition trade raw performance for manageability; others narrow their focus substantially to address performance concerns, e.g., Soukup's data structure composition techniques [15]. Thus, while composition approaches are useful in many applications, they are rarely a complete solution.

3.2. Generation-Based Systems

Generation-based systems systems are basically programs that build other programs. They codify the important decisions and tradeoffs that underly the construction of the particular class of applications they target: architecture, algorithm, data structure, optimization, etc. Biggerstaff and Perlis distinguish three categories of generation-based systems [3].

Language-based systems compile a highly abstracted language description (which may or may not be executable by itself) into low-level code, hiding as many implementation decisions as possible in the compiler itself. An important example of such a language is SETL [7, 14], which specifies computations in terms of abstract sets only, and is devoid any data structure details in the conventional sense. That is, the critical computations are specified, but the details of the storage mechanisms are removed. For example, all computations could be specified using only sets, and simple set operations, in place of detailed data structures; this is done in SETL [14, 7] and REFINE^{TM}, a commercial object-oriented programming environment. This focuses attention on the fundamental computation, while freeing the specification of side effects resulting from the choice of specific structures like stacks or queues, heaps or trees. SETL emphasizes abstract data structure types like sets and maps. Data structure implementation decisions are delayed into the SETL compiler.

The SETL compiler selects data structure implementations by examining a detailed description of structure relationships implied by the SETL language description. Selection of appropriate data structures is necessary before program construction can begin [8, 13]. This is especially true when the input specification is in the form of a very high-level language devoid of specific data structure choices, such as in the case of SETL. The automatic selection of data structures requires information on the intended usage of particular data structures. Examples of this kind of information are required access times, maximum space allocation and data structure assignment inter-relationships.

Data structure selection is often limited by the small library of data structures available to choose from, for any particular needed structure. Implementation issues include the use of domain knowledge in the selection criteria, and the quality of the data structure library. SETL is capable of producing a wide variety of complex data structure implementations, for example, data structures requiring indexing operations or bit operations, from the detailed description of their structural relationship.

A related example of a system employing data structure selection techniques is that of Low [8]. This work provided selection from among a fixed library of implementations. The selection library was basic but included the concept of sets. The selection criteria was inflexible, and was always fixed to minimize the product of memory usage and execution time. Application domain knowledge was not included. Low used a dynamic run-time analysis of the program to derive the needed execution time statistics. The search technique was a form of hill climbing, using the selection criteria as the evaluation heuristic. An n-tuple representation of current selections for each structure was used. The hill climbing approach varied one entity selection at a time noting changes to the selection criteria. Maintaining all entity selections and their relative merit during the search process facilitates any necessary backtrack. A useful contribution of this research was the use of execution statistics as selection criteria and the n-tuple representation form.

Application-generators literally embed, as code, all the relevant decisions necessary to design a particular piece of software from a specification for its performance. The analogy to VLSI module generators, for PLAs, RAMs, datapaths, etc., from silicon compiler systems is most apt here. The important point about such generators is how heavily they rely on what is usually referred to as *domain knowledge*, i.e., expertise about how to build a particular software application. Despite the seductive appeal of using composition-based ideas to snap together abstract building blocks and arrive at complete systems, it has been observed [3, 9] that most nontrivial applications devote the bulk of their code to implementing this sort of idiosyncratic expertise.

An example of applying application domain knowledge in automatic programming is the DRACO system [9]. The DRACO system constructs a program using re-usable program parts. Programs are derived from a domain-language specification using the standard compiler technologies of grammers and parsers. The problem of writing software is thus transferred from writing the implementation code to correctly formulating the domain. Formulation requires a fairly substantial domain analysis and user investment. The domain-language specification is a difficult design task. The domain-language description embodies two separate problems: acquiring the domain knowledge to be encapsulated as tactics and formulating these tactics into a format suitable for input to a parser such as BNF.

Finally, **Transformation-based** systems strive to implement a sequence of transformations that incrementally evolve a high-level specification into executable code. Transformations may be automatic, or available as a suite of options to be guided by a software designer. Transformation approaches have been used successfully in many automatic programming applications [1, 2, 5, 6]. An example of a system using the program transformation approach is the PSI/SYN system [1, 6]. The input to PSI is a program description in a subset of English. The output is a LISP implementation. PSI/SYN incorporates a system-level approach to select data structures and algorithms with the mechanics of program transformation. LIBRA chooses data structure representations by first identifying all plausible options, and then searching for the some "best" implementation from the plausible option space. Efficiency analysis of the remaining data structure representations in the context of the entire program controls and guides search. Efficiency analysis can include domain-specific knowledge. The program under development is refined in stages to further control and limit search decisions.

3.3. Chapter Summary

This chapter has provided a rather brief overview of techniques developed in automatic program synthesis, such as code generation, data structure selection, and algorithm selection. The goal of these techniques is to develop and support a completely automatic program generator including automatic algorithm selection. Ideally, an automatic program generator should vastly simplify program development tasks by avoiding tedious manual tasks such as coding and implementation debugging. Present technology falls short of this goal. However, our argument is that in some well-understood, bounded application domains, this synthesis technology is actually quite useful. The next chapter shows how we have integrated some of these ideas into a working synthesis architecture, called ELF, targeted at the specific problem of synthesizing maze routing software.

References

[1] D. Barstow, "An Experiment in Knowledge-based Automatic Programming", *Artificial Intelligence*, Vol. 12 1979, pp. 73-119.

[2] D. Barstow, "A Perspective on Automatic Programming", *AI Magazine*, Vol. 5 1984, pp. 5-27.

[3] T.J. Biggerstaff and A.J. Perlis, editors, *Software Reusability*, Addison-Wesley Publishing Company, Vol. 1, 1989.

[4] H. Brown, C. Tong, and G. Foyster, "Palladio: An Exploratory Environment for Circuit Design", *Computer*, December 1983, pp. 41-56.

[5] M. Broy and P. Pepper, "Program Development as a Formal Activity", *IEEE Transactions on Software Engineering*, Vol. 7, Jan 1981.

[6] E. Kant, "On the Efficient Synthesis of Efficient Programs", *Artificial Intelligence 20* 1983.

[7] S. Liu, *Auotmatic Data Structure Choices in SETL*, Courant Institute of Mathematical Sciences, 1979.

[8] J. R. Low, "Automatic Data Structure Selection: An example and overview", *Communications of the ACM*, ACM, May 1978, pp. 376-384.

[9] J. Neighbors, "The DRACO Approach to Constructing Software from Reusable Components", *Readings in Artificial Intelligence and Software Engineering* 1986.

[10] R. Newton, "A D&T Roundtable: CAD Software Development", *IEEE Design and Test of Computers*, August 1988.

[11] C.V. Ramamoorthy, "Object-Oriented Systems", *IEEE Expert*, Fall 1988.

[12] C. Rich and R.C. Waters, editors, *Readings in Artificial Intelligence and Software Engineering*, Morgan Kaufmann Publishers, Inc., 1986.

[13] L. Rowe and F. Tonge, "Automating the Selection of Implementation

Structures'', *Readings in Artificial Intelligence and Software Engineering* 1986.

[14] E. Schonberg, J. Schwartz and M. Sharir, ''An Automatic Technique for Selection of Data Representations in SETL Programs'', *ACM Transactions on Programming Languages and Systems*, Vol. 3, April 1981, pp. 126-143.

[15] J. Soukup, ''VLSI Databases, Frameworks, and Open Systems''.

Chapter 4
ELF: A Program Synthesis Architecture

ELF ia a program architecture: an organization of synthesis strategies aimed at transforming an abstract, non-executable description of a software tool into real, working software. Although ELF relies on many program synthesis ideas, the real focus of the ELF prototypes is to illustrate the feasibility of integrating domain knowledge and program synthesis knowledge for the automatic generation of real-world application tools. Maze routers are our target application domain. However, it is important to understand that ELF is a prototype. We do not expect to be able to generate a perfectly tuned, perfectly customized, high-performance router; rather our intent is to demonstrate the feasibility of automatically generating tools. Hence, we intend to focus on the construction of "reasonably functional and efficient" routers from only a high-level description of the routing task and from domain knowledge. As we shall see in subsequent chapters, we measure success by our ability to generate and demonstrate such routers, and to retarget the ELF generator to different maze router variants. Producing routers over a variety of technologies illustrates how this architecture copes with shifting technology environments. This chapter introduces the major components of the ELF architecture, and the basic synthesis strategies and representation schemes used by ELF.

4.1. Combining Router Knowledge with Program Synthesis Knowledge

ELF is a generation-based system that has aspects of the language-based, application-generator and transformation-based approaches discussed in the previous chapter. ELF combines a very high level language (actually resembling SETL) for specification of abstract design styles, codified knowledge about router mechanics and coding mechanics, and a transformational approach to render an automatically derived, non-executable router architecture into actual executable code. ELF is novel in large part because of its scale: it integrates a variety of program synthesis techniques, e.g., algorithm selection [10], data structure selection [9, 11], program transformation including data structure selection [1, 8], and the use of domain knowledge within the transformation process [2] into a working system targeted at a practical, large-scale CAD application.

Before proceeding to describe the ELF router generator, it is worth addressing one obvious criticism of our generation-based approach. This is the assertion that the reason there are many different maze router implementations is not that they are necessary, but rather, that we do not have an adequate model for a universal maze router tool. Given the hundreds of variants in the literature, and in industrial use, and the continuing creation of tuned special-purpose maze routers (e.g., for gate arrays [6], PCBs [5], even sea of gates ICs [7]), we are skeptical that the size and complexity of such a tool would be manageable. Nevertheless, the point is not whether such a universally retargetable tool could be constructed, but rather, *how* any retargetable tool should be built. The aim of our research on generation-based approaches for CAD tools is to understand how knowledge about re-engineering large, complex CAD software can be codified so as to mitigate the well-known problems of maintaining and modifying such software.

Our architecture is composed of three separate stages, as illustrated in Figure 4-1. The first stage, *input*, reads all routing technology specifications and physical constraints from the user, including a set of very high-level router algorithm descriptions. The second stage, *selection*, interprets routing requirements to select the appropriate data structures and algorithms to complete the task. Examples of knowledge sources used in selection include domain-specific router knowledge, knowledge of data structure capabilities, and algorithm performance knowledge. The third stage, *code generation*, uses the algorithm and data structure choices from the selection section, a set of very high-level router algorithm descriptions, as well as additional router domain-specific knowledge to generate executable output code using program transformation techniques. Representation and interaction issues that must be addressed in an implementation of this architecture include:

Input Stage
- Representation of routing algorithms
- Representation of algorithm, application, and fabrication constraints

Selection Stage
- Representation of and role of domain knowledge
- Representation of data structures
- Interaction (or Separation) of algorithm and data structure selection
- Selection tradeoffs

Output Code Generator Stage
- Intermediate representation for synthesized code

- Code elaboration and generation

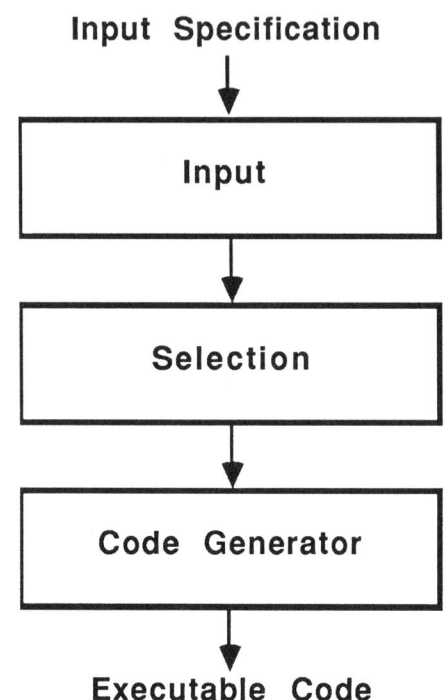

Figure 4-1: Router Generator Architecture

Our implementation of this architecture, called ELF, will be referred to briefly in this chapter, and in greater detail in subsequent chapters. ELF produces executable routers in the C programming language. The following sections outline our approach to handling these architecture issues along with supporting rationale. We first discuss knowledge representation issues, then discuss each stage of the architecture separately.

4.2. Algorithm Schema Representation

We rely on a very high-level language specification for the routing algorithms. These specifications are devoid of *any* specific implementation decisions; in particular, they specify no particular data structures. Our algorithm specifications are represented in a custom-designed language called ADL[1], which is essentially a much simplified variant of SETL [11]. ADL's syntax is program-like in contrast to SETL's set symbolic mathematical syntax. Program constructs in ADL are English-language words, rather than set symbols. Each ADL description specifies a *schema* for the algorithm it represents.

The ADL language adopts the C language's variable scope definition. For example, a variable declared in the local segment of a function that specifies variable data structure interrelationships is assumed to be local to that function. Parameters in ADL are call by value unless modified internally by the function. In the latter case, ELF generates a call by reference parameter passing protocol. This will be covered in more detail in the code generation chapter.

An ADL description consists of two segments. The first segment details the data structure interrelationships of the local variables. These interrelationships are not limited to be within the same variable scope definition--indeed the most useful interrelationships are those from local to global. For example in Figure 4-2, if a locally defined data structure is related to another globally defined data structure, then that interrelationship is specified in the first segment of the ADL description. This assists the data structure selection for the locally defined data

---

[1] Algorithm Development Language.

structure, and the algorithmic timing estimation and code synthesis. Interrelationships can be defined by dimensions[2] between data structures, by the number of dimensions in the dependent data structure, or by data structure implementation. In Figure 4-2, parameter `so_easy` maps to global variable `global_variable` so it therefore must vary throughout the range of `global_variable`. Local variable `local_variable` specifies elements of `global_variable` so the number of dimensions in the dependent data structure `global_variable` is required for full specification. The keyword CODE signals the end of this interrelationship definition segment and the beginning of the second segment.

```
SUBROUTINE adl
PARAMETER so_easy
  VARY_BY global_variable END

SET local_variable
  DIMENSIONED_BY global_variable END

CODE
```

Figure 4-2: ADL description: Interrelationship Segment

Figure 4-3 is a sample ADL description of a router subtask. This particular description is of the expansion router subtask. The procedure looks at each element within the set `wavefront` and references it through `same_cost`.

---

[2] A *dimension* of a data structure is defined as any method by which the elements of the data structure may be organized (i.e., if a data structure is only organized by one method, say its elements are increasing in value as in the result of a sorting algorithm, then the set is one dimensional). Set ordering may be imposed by the use of some knowledge source.

Each element of the same_cost within the set wavefront is sequentially chosen for expansion, in the second LOOP statement. The Test statement checks each expandable site for validity, e.g., that it has not already been expanded, and that it is within some router-defined limits, either hard physical limits like the edges of a circuit board, or soft limits set up to reduce the expansion search space. If the expandable site passes the validity test, then we check to see if we have reached the target in check_target which is defined elsewhere, and compute the cost of expanding into this cell. The new cell is then put back into the data structure wavefront along with all other elements with the same cost value, and the old cell is removed from the same data structure. The first ENDLOOP, corresponding to those cells at the same cost within wavefront, ensures all sites are expanded. The final ENDLOOP, corresponding to each wavefront element, ensures all wavefront elements are available for expansion.

```
LOOP path_not_found IS TRUE END
   LOOP site IN same_cost OF wavefront END
      TEST site NOT expanded AND
           site IN limits AND
           path_not_found IS TRUE END
         ASSIGN path_not_found TO
            check_target WITH site END
         ASSIGN cost TO cost + cost_function END
         PUT site IN wavefront OF cells WITH cost END
         REMOVE site FROM
            wavefront OF cells WITH same_cost END
      ENDTEST
   ENDLOOP
ENDLOOP
```

Figure 4-3: ADL Algorithm Specification Example

Note that this sample ADL description only specifies basic computations independent of implementation constraints (such as data structure selection) using sets and set operations to reference all complex data objects. ADL does not coerce these sets into specific implementations. For example, in Figure 4-3, the set wavefront is only known to be accessed via same_cost. In the actual implementation, wavefront may be implemented as a multi-dimensional array or a linked-list of lists. Hence, same_cost may then be chosen to be implemented as a multi-dimensional array index that is linked to the chosen dimensions of wavefront, or as a complex pointer set or pointer record structure.

Note also that the ADL fragment does not specify *how* same_cost is ordered nor does it specify *how* to generate each site from the same_cost variable. These actions require router domain knowledge. The formulation of the expansion subtask description did not require specific router domain knowledge.

4.3. Data Structure Style Representation

The representations employed for data structures are based heavily on the representations used in typical transformation approaches to program synthesis [8]. Information about specific component and subcomponent implementations, for example, access time characteristics and space requirements, is necessary. In short, data structure *efficiency* information is required. In addition, we store information about data structure *suitabilities*, i.e., router expertise about the ability of a particular data structure candidate to cope with a particular algorithm selection or technology constraints.

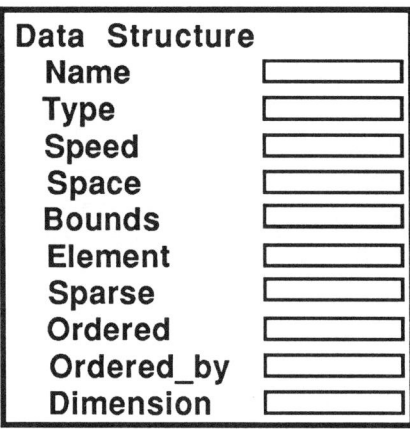

Figure 4-4: Data Structure Representation

Data structures in ELF are represented in a simple object-oriented fashion. Templates for elementary structures such as one-dimensional arrays and lists are supported (see Figure 4-4), with attribute slots for obvious characteristics such as size and access-time. In addition, composition of structures is permitted using a simple part/whole attribute, so that we can construct arrays of lists, or trees of arrays, and so forth. We create a composition of these representations for each necessary data structure. This construction process is termed *selection* and will be discussed in the Selection Stage. For simplicity, we refer to each layer of such a hierarchically composed data structure as a *dimension* of that data structure. Each dimension contains domain knowledge of the expected behavior within the user specified domain (e.g., set sparsity, set ordering, indices for accessing ordered structures). An example of a representation of a multi-dimensional data structure is shown in Figure 4-5.

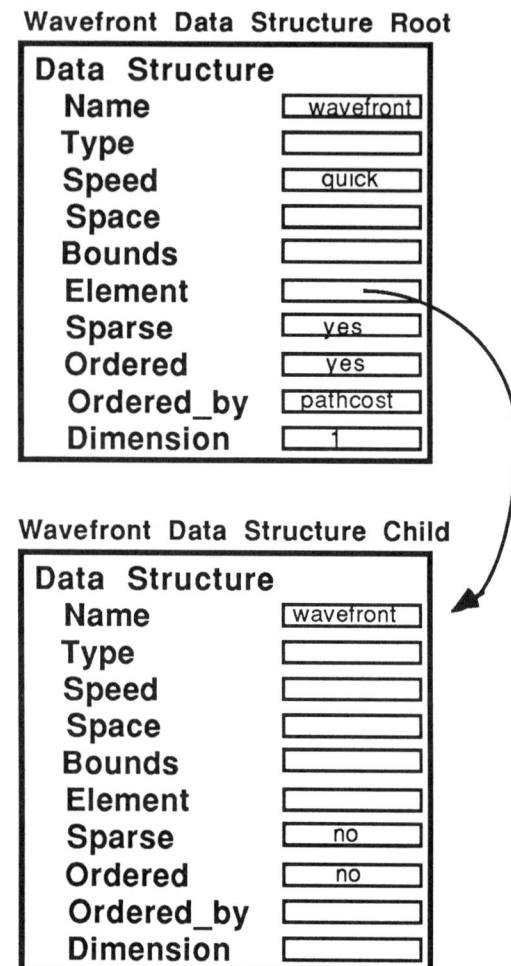

Figure 4-5: Multi-dimensional Data Structure Representation

In this example, the first dimension has characteristics such as quick-access time and relative sparseness. The second dimension has different characteristics: the set entries are not relatively sparse and each element is related to another separate data structure. These characteristics do not exist in a void, but rather are a direct result of some application of domain-specific knowledge. The method by which domain-specific knowledge is applied to produce these characteristics is described in Chapter 6.

4.4. Intermediate Representation for Synthesized Code

We chose to employ an intermediate representation for the synthesized code. That is, algorithm selection and data structure selection deal with objects that are not actual fragments of code, but a higher-level, more abstract intermediate representation for code. The code generator then transforms this intermediate representation down to final executable code. An intermediate representation allows for varying depths of transformations that can occur in code generation. For example, during code generation the recognition of a loop construct in the intermediate representation requires further refinement to choose an actual implementation of the loop. In addition, an intermediate representation allows code generation and elaboration (which we discuss later) to be treated as a refinement step. The intermediate representation is a syntax tree, much like that found in conventional compilers. An example is shown in Figure 4-6. Initially, the representation is just ADL text. (In contrast to other languages, the value of the right hand side of the ASSIGN statement is assigned to the left hand side.) ADL is then transformed into the intermediate tree-like representation first by applying semantic knowledge, or domain-specific program synthesis knowledge. Program synthesis knowledge then refines the intermediate representation into text in an executable form. This method is ideal for the type of expansion required from any necessary refinement step.

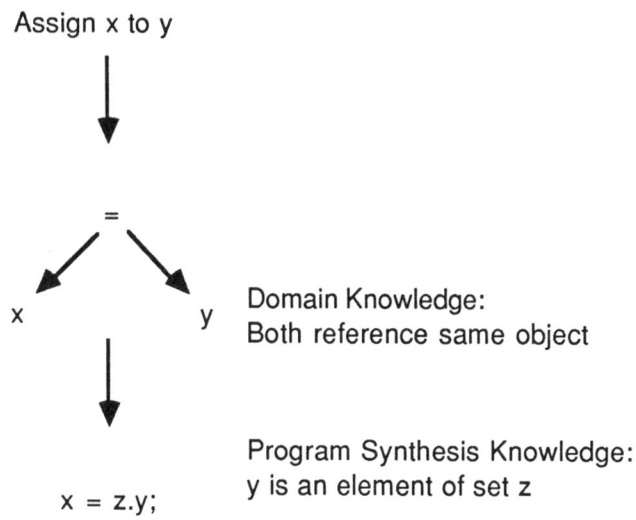

Figure 4-6: Use of Knowledge in Intermediate Representations

4.5. Domain Knowledge Representation Using a Rule-Based System

A rule-based implementation is appropriate in programming situations where the domain information can be simply expressed as a set of situation-action pairs, and there is an expectation that the codified information will grow as the system is developed. Router domain knowledge is best represented as sets of situation-action pairs. For example, domain knowledge about the effects of basic routing requirements on data structure implementation choices, and the effects of data structure design on access time/space tradeoffs all can be expressed as situation-action rule pairs.

The utility of situation-action rules is even seen, albeit rather implicitly, in the algorithmic routers discussed in Chapter 2. In these cases, information about different router variants is represented as complicated sets of "case-like" statements in the code implementing the router. Our experience suggests that expertise about router construction can be naturally codified as rules relating data structures and algorithm choices. A rule-based implementation thus has the advantage that, rather than being hidden as in a conventional algorithmic implementation, the interrelationships between algorithms and data structures are explicit in the rules comprising the generator. ELF is therefore implemented as a rule-based system.

OPS5 [3] was chosen for its accessibility, and for its ease in generating rules. This ease allowed us to build a prototype quickly. OPS83 has become available since we began development of ELF, and in the future we could convert to that implementation language. In addition, there are some portions of ELF implemented directly in LISP, but these are very low-level support functions. The essential components of the ELF implementation are in OPS5.

Knowledge may be represented as objects, e.g. ADL algorithm representations and data structure characteristic representations, or as a set of rules within the ELF system. Informally, we can identify the following classes of knowledge used in ELF:

1. **Design Generation knowledge**
    a. Router structure knowledge
    b. Routing phase requirement knowledge
    c. Router dependency knowledge
2. **Program Synthesis knowledge**
    a. Application language syntactic knowledge

      b. Data structure implementation knowledge
  3. **Design Interaction knowledge**
  4. **ELF-control knowledge**

All of these types of knowledge are represented as rules within the ELF system. In this section, we outline and describe each type of knowledge used in the ELF system.

4.5.1. Design Generation Knowledge

More information is required to write a router than just a very high-level ADL description and some information about data structure possibilities. This required information is what we refer to as the design generation knowledge, i.e., expertise about the implementation of maze routers. Design generation knowledge is used to guide the selections made in the selection stage, as well as to guide code generation. Design generation knowledge is not represented as an exhaustive set of all combinations of data structure and algorithm selections and their ramifications, but rather as the ramifications of certain critical sets of combinations. This design generation knowledge representation changes the domain knowledge from a many-to-one style to a many-to-many style.

4.5.1.1. Router Structure Knowledge

Generally, the first task of the router generator is to deduce the routing design demands, for example, the criteria of just what constitutes a solution for the application routing problem as represented by the four routing phases (setup, expansion, backtrace and cleanup). Primarily, this knowledge is of the capabilities and behaviors of the four routing phases. Each behavior is represented as a task, with subtasks within its phases. For example, the *setup* phase tasks are: take as input a netlist, select a net for routing, and output the net to the *expansion* phase. Subtasks in the setup phase include internal net ordering and internal terminal ordering. For example, a PCB wire-router striving to minimize the effects of congestion might attempt to route the shortest wires first, as they cause the least blockage problems. This knowledge implies that the input netlist arrangement must be sorted by estimated length, and therefore included in the routing solution, before being submitted to the *expansion* phase. Router structure knowledge also affects the selection of expansion search and backtrace algorithm subtasks within the *expansion* and *backtrace* phases respectively.

4.5.1.2. Routing Phase Requirement Knowledge

Once the framework of the solution is deduced (i.e., the router's infrastructure is composed of phases, tasks, and subtasks), phase requirement knowledge is needed to generate a more complete description of the desired router design. Phase requirement knowledge differs from router structure knowledge in that router structure knowledge is concerned with generating solution phases, tasks and subtasks, while phase requirement knowledge is concerned with how to *fulfill* these requirements in ADL. Examples of phase requirement knowledge range from modifications to the expansion search or backtrace algorithms, to identifying suitable algorithms for each task or subtask.

4.5.1.3. Router Dependency Knowledge

It is difficult to make decisions on phase composition and implementation without knowledge of how particular decisions affect other decisions. This type of knowledge is called router dependency knowledge. These interaction rules cover any search, or evaluation within a search, where multiple design decisions are required. At the most abstract level, router dependency knowledge is required to represent the interrelationship between phases. For example, if memory requirements are to be minimized, then the embedded backtrace pointer might be field- or bit-mapped by the expansion phase. The backtrace phase must then be capable of translating the possibly obscure pointer representation to some usable (within the backtrace phase) format. As the design is refined, router dependency knowledge is required to match the search algorithm's representation to the physical constraints of the problem. As an example of this, consider the impact of allowing 45 degree wiring in addition to Manhattan style wiring. This has an obvious effect on the expansion phase of a router, with algorithm changes corresponding to more complex adjacency rules. Router dependency knowledge matches the algorithmic representation of the expand task within the expansion phase to the adjacency rules by guiding code generation to include code to expand a cell along 45 degree as well as 90 degree expansion paths.

4.5.2. Program Synthesis Knowledge

Program synthesis knowledge guides the formulation of an efficient executable implementation from the abstract representations of task, subtasks, and high-level algorithm representations. Information on the executable implementation is what we refer to as program synthesis knowledge. There are two major examples of this expertise, namely, how data structures are declared in the coding domain and how to produce the access code for the abstract representations. Program synthesis knowledge is used to guide selections made in the selection stage, as well as to guide code generation.

4.5.2.1. Application Language Syntactic Knowledge

The application language imposes another set of constraints on the routing problem. Not only must a router be refined to all routing tasks and subtasks, but the subtasks must also be transformed into an executable, syntactically correct, target language. Application language syntactic knowledge is the coding expert knowledge source within the ELF system. This knowledge encompasses the C language (the ELF output executable language) syntax requirements, and how the language constructs impact performance. A simple example is the set of constructs that form a loop, a *for*, a *while-do*, or a *do-until*. Each of these constructs has specific syntactic requirements, and while they are semantically interchangeable either through the introduction of new variables or code repetition, each is suited towards specific implementation usages. How to select the appropriate loop constructs for varying implementation conditions provides an example of this type of knowledge.

4.5.2.2. Data Structure Implementation Knowledge

Data structure implementation knowledge is knowledge of data structure efficiency characteristics for time and space within the coding domain. Of course, it is dependent upon the target application language. For example, given an assignment statement in the C application language, this knowledge is used to select the appropriate data structure access code (e.g., the left hand side of the assignment statement), or to evaluate the access characteristics (e.g., speed) of the data structure implementation. These access characteristics are used to select and refine data structure implementations in the selection stage. In addition, this information guides code generation by using instances of known access protocols (e.g., how to access an array, a list, an element within a record, etc.) to resolve complex data structure access requirements within the syntactical requirements of the C application language.

4.5.3. Domain Interaction Knowledge

Router domain knowledge and program synthesis knowledge often interact while attempting to achieve a common goal. This is particularly evident while selecting abstract representations of the data structures and algorithms. Code generation, interestingly enough, also evidences domain knowledge and program synthesis knowledge interaction. Interactions between domain knowledge and program synthesis knowledge are mostly due to the inherent independencies among data structure implementations within an algorithmic structure. These independencies are represented as domain interaction knowledge.

An example of a design activity displaying design interaction knowledge is algorithm timing analysis. This timing approximation becomes more accurate as information on the data structure implementation becomes more concrete. Algorithms are stressed by the chosen data structure implementation. For example, the execution time of a sorting algorithm (an example of domain-provided knowledge) is highly dependent upon the data structure to be sorted; a relatively flat data structure is generally faster (e.g., an array or list), than a complex hierarchical data structure (e.g., a tree of lists of arrays) when the hierarchy indexing element does not correspond to the sorting element. The impact of data structure implementation on algorithm operation time is termed *algorithm stress knowledge.* Algorithm stress knowledge is a type of domain interaction knowledge. Instances of this knowledge include the approximate timing requirements of an algorithm given alternative data structure implementations, as well as the effects of design tradeoffs between algorithm and data structure selection. As an example of this, again consider the impact of allowing 45 degree wiring in addition to Manhattan wiring. This has an obvious effect on all data structures having to do with the wavefront expansion: they must modify the definition of routing adjacencies and availabilities. The expanded routing adjacencies affect the timing requirements of the expansion search algorithm. Algorithm stress knowledge guides the formulation of these timing estimates using data structure implementation knowledge.

Another example of a design activity displaying similar design interaction knowledge is the data structure analysis dealing with algorithm requirements. This is termed *data structure stress knowledge.* This knowledge concerns fairly direct requirements for data structure choices for routing. An example is the elementary knowledge that for a detailed maze router there must be a data structure for the routing grid, for the expanding wavefront, and so forth. A more complex example is the knowledge used to design data structures. Using the same 45 degree wiring addition as above, the wavefront data structure and the routing algorithms must be modified to accommodate this expanded routing adjacency definition. The wavefront data structure is stressed by the 45 degree wiring addition and must be modified to successfully represent the problem.

4.5.4. ELF-control Knowledge

The resolution of when to make particular routing task decisions is provided within ELF's problem-solving architecture. This architecture involves the scheduling of all identified task and subtask designs and is called ELF-control knowledge. This knowledge prioritizes tasks and subtasks and recognizes which particular subtask is needed to resolve a particular behavior. For example, this knowledge recognizes when it is necessary to carry out the subtask to sort the input netlist, or when it is necessary to carry out the final coding subtask.

4.6. Architecture Overview

ELF is a knowledge-based synthesis architecture targeted towards the generation of routing software. ELF integrates knowledge of the routing domain with knowledge of generic programming mechanics. This section provides a high-level overview of the three stages--input, selection and code generation--of the ELF synthesis architecture as shown in Figure 4-7. More detailed descriptions of each stage, including the interaction of rules and knowledge sources, are provided in Chapters 5, 6, and 7.

4.6.1. Input Stage

This stage reads from the user all routing task specifications and physical constraints for use by the selection stage. The most important routing task specification is the routing algorithm specification. These specifications define the order and kinds of computations, i.e., the routing tasks and subtasks, required to complete the routing problem for the different types of routers.

Automatic Programming Applied to VLSI CAD Software: A Case Study   63

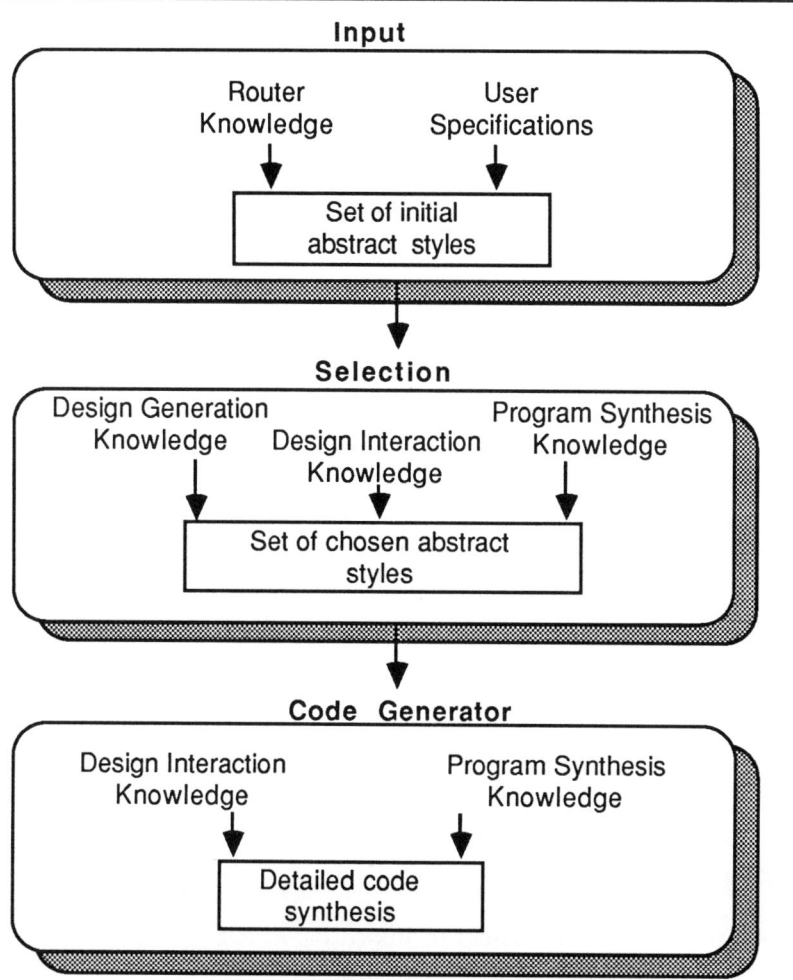

Figure 4-7: ELF Synthesis Architecture

4.6.2. Selection Stage

The Selection stage chooses the appropriate data structures and algorithms to complete the desired routing task. Decision-making requires information on such topics as data structure capabilities and algorithm suitability. Domain knowledge guides data structure and algorithm selection. Domain knowledge includes such information as: data structure selection guidelines based upon the set of technology specifications given in the Input stage, and the basic algorithm requirements known to be necessary to complete the desired routing task. The representation of domain knowledge and data structure information are an important part of this stage. In addition, the methods for selecting data structures and algorithms are important issues here. Finally, the *separation* of data structure and algorithm selections--a design decision for our synthesis architecture--has consequences for the organization of the overall synthesis task. We discuss each of these issues later in this section.

4.6.3. Selection

Selection is heavily domain-knowledge dependent. Information for specific implementations, for example, access time characteristics and space requirements for data structures and time characteristics and data structure usage for algorithms, are required. In addition, we employ information such as data structure "suitabilities", i.e., router expertise about the ability of a particular data structure to support a particular algorithm selection or technology constraint, within the routing domain. Similarly, we employ information on algorithm "suitabilities", i.e., router expertise about the ability of a particular algorithm to support the operational demands of the router phase and the candidate data structures. For each data structure in the desired routing task, such as the grid data structure, wavefront structure, etc., a specific implementation must be chosen based on the needs of the particular algorithm candidates. For example, a backtrace data structure implementation for a depth-first maze router has different time/space characteristics than does a breadth-first maze router. An algorithm selection for wavefront expansion affects the implementation chosen for the backtrace data structure. This illustrates the importance of characterizing and encompassing the interactions and interdependencies between data structure and algorithm selections.

4.6.4. Separation of Algorithm and Data Structure Selection

An important design decision for our synthesis architecture is the *separation* of algorithm selection and data structure selection. Abstract algorithm and data structure representations are often used in program transformation. Some algorithms are specified without reference to specific data structure implementations and possible data structure implementations are refined to match the requirements. In addition, the algorithms themselves are data structure implementation dependent. The separation of algorithm and data structure synthesis follows naturally from transformation systems, which define algorithms independently of any specific data structure implementation [4]. Within the Selection Stage, algorithm selection is performed *separate* from data structure selection, but the ramifications of a specific selection in one affect the other [12]. In our synthesis architecture, illustrated in Figure 4-8, algorithm selection is based on problem description input, and router domain knowledge. Data structure selection is based on problem description input, router domain knowledge, and implementation quantifiers, such as speed and size, that are dependent on the currently selected algorithms' use of specific data structures.

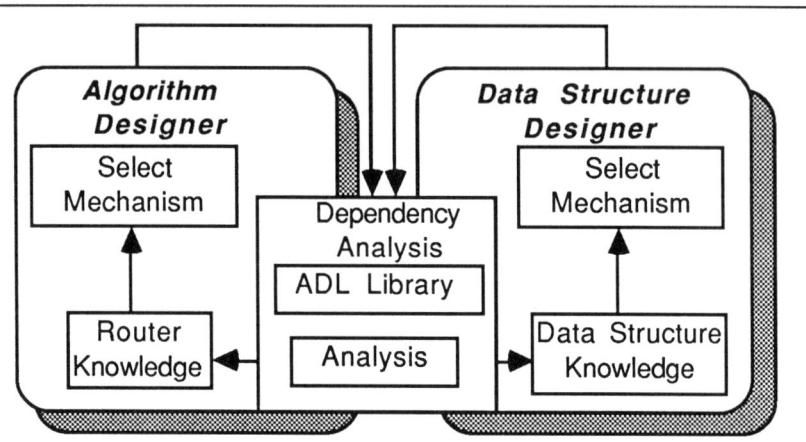

Figure 4-8: Data Structure and Algorithm Designer Interrelationship

4.6.5. Output Code Generator Stage

The output code generator stage generates the required output code for the router using router domain knowledge, a set of selected very high-level algorithm descriptions, and a set of data structure implementations. As in other systems employing the program transformation approach, we must consider issues of the depth and breadth of a single transformation. Like the PSI/SYN [1, 8] system, the code generator stage generates final code in a sequence of small transformations. The process starts from ADL, progresses to an intermediate representation for the code to be synthesized, then finishes with the executable code. Code elaboration uses router domain knowledge, and data structure and output code interrelationships to translate the very high-level description to the desired final format.

4.7. Architecture Overview

ELF Stage and Input	Produces	Using Knowledge Source
Input: User Description	Set of multiple candidates for each task and Subtask	Router Domain Router Structure Routing Phase Requirement
	Task set Subtask set	Routing Phase Requirement
Selection: Set of Multiple Candidates	Set of selected candidates for each task and subtask	Design Interaction Algorithm Stress Data Structure Stress Router Structure Router Domain Data Structure Implement.
	Performance Information	Algorithm Stress Design Interaction Router Structure
Code Generation: Set of selected candidates for each task and subtask	Intermediate Code Representation Final output Code	Data Structure Implement. Router Structure Application Lang. Syntactic Design Interaction

Figure 4-9: ELF Architecture Knowledge Characterization

Figure 4-9 summarizes the actions and results of each stage, and the knowledge used to produce each object within each stage. The Input stage begins with the user input specifications and produces implementation candidates using design generation knowledge. Several candidates for each routing task and subtask may be produced. The Selection stage begins with these candidates and selects the appropriate candidate using design generation knowledge, program synthesis knowledge, and design interaction knowledge. Candidate representations, for example, ADL algorithm representations and data structure representations, are analyzed and modified to reflect selection decisions. The Code Generation stage begins with the selected candidates and produces first an intermediate code representation, then the final code using program synthesis knowledge, and design interaction knowledge.

4.8. Chapter Summary

This chapter introduced the ELF synthesis architecture. The use of program transformation, combined with router domain knowledge, was justified as both appropriate and desirable for this research. Issues central to this approach, such as the role of domain knowledge, algorithm and data structure selection, and algorithm and data structure representation were discussed. Finally, this chapter briefly discussed the high-level operation of ELF, our router generator, a prototype designed to test the feasibility of this architecture. In the following three chapters, we revisit the major components of the ELF synthesis architecture, and describe their functions in more complete detail.

References

[1] D. Barstow, "An Experiment in Knowledge-based Automatic Programming", *Artificial Intelligence*, Vol. 12 1979, pp. 73-119.

[2] D. Barstow, "A Perspective on Automatic Programming", *AI Magazine*, Vol. 5 1984, pp. 5-27.

[3] L. Brownston, R. Farrell, E. Kant, and N. Martin, *Programming Expert Systems in OPS5*, Addison-Wesley, 1985.

[4] N. Dershowitz, "Program Abstraction and Instantiation", *ACM Transactions on Programming Languages and Systems*, Vol. 7, July 1985, pp. 446-477.

[5] R.J. Enbody and J.C. Du, "Near-Optimal N-Layer Chanel Routing", *23rd Design Automation Conference*, 1986, pp. 708-713.

[6] D. Hightower, "The Lee Router Revisited", *ICCAD-83*, IEEE, 1983.

[7] M. Igusa, M. Beardslee, and A. Sangiovanni-Vincentelli, "ORCA A Sea-of-Gates Place and Route System", *26th Design Automation Conference*, 1989, pp. 122-127.

[8] E. Kant, "On the Efficient Synthesis of Efficient Programs", *Artificial Intelligence 20* 1983.

[9] S. Liu, *Auotmatic Data Structure Choices in SETL*, Courant Institute of Mathematical Sciences, 1979.

[10] Robert McCartney, "Synthesizing Algorithms with Performance Constraints", Tech. report, Brown University, December 1987.

[11] E. Schonberg, J. Schwartz and M. Sharir, "An Automatic Technique for Selection of Data Representations in SETL Programs", *ACM Transactions on Programming Languages and Systems*, Vol. 3, April 1981, pp. 126-143.

[12] N. Wirth, "Program development by stepwise refinement", *Communications of the ACM*, ACM, April 1971.

Chapter 5
The Input Stage

The Input Stage takes as input a user specification template consisting of a set of technology constraints and produces the problem description used in the Selection Stage. This problem description consists of an ordered set of tasks and subtasks that define the router design process. For each task, the Input Stage generates a set of candidate solution options. These options correspond to ADL algorithm representations or to data structure representations. Options are generated by applying router domain knowledge to the user specification. This chapter discusses some operational issues, the types of knowledge used, and examples of typical rules in the Input Stage.

5.1. Input Stage Operation

The Input Stage is the user-interface module in the ELF system. The user specifies the desired router application by a set of technology constraints. These technology constraints encompasses the following: algorithm schema alternatives, application requirements, and fabrication specifications. The set of technology constraints does not have to be complete. ELF contains two inference mechanisms in this stage. One inference mechanism infers any necessary technology constraints not detailed in the user specification. This mechanism is guided by both *router dependency* and *router structure knowledge*. In addition, this stage contains a second inference mechanism to map each user specification onto the relevant set of router tasks and subtasks. This mechanism is mainly guided by *router structure knowledge*.

The user specification takes the form of a template. Figure 5-1 is a template showing the available parameter options for each user specification. Only those specifications required to describe a given routing application are entered into the template. For example, to emphasize minimum routing execution time at the cost of possibly increased space consumption, the user would fill in "time" in the `minimize` entry in the template. Again, not all entries are required to be specified for ELF to correctly synthesize a router. In fact, in most cases, the majority of the entries are left blank, leaving ELF to deduce reasonable choices for the remaining specifications using router domain knowledge. Figure 5-2 shows the types of domain knowledge used in this stage. The Y axis, labeled Percent of Knowledge Used in Stage, refers to the fraction of rules in this stage that involve that particular type of knowledge. There are four types of knowledge sources.

In addition, the set of technology constraints does not represent a fully-constrained routing problem. Design decisions, represented by tasks and subtasks, are still necessary even if ELF is given a complete specification set. The most obvious example is the lack of any data structure implementation selection within the specification set. A more subtle example is the lack of access to individual low-level design decisions from the user-interface. ELF takes a one-to-many approach to design decisions. From one high-level

```
algorithm
  net_sorting              shortest_net_first,
                           longest_net_first, no
  node_sorting             yes, no
  cost_function
    association            cell, boundary, both
    congestion_multiplier  #
    via_multiplier         #
    layer_multiplier       #
    penalty                linear, exponential, step
  input                    descriptive_netlist_file
  output                   descriptive_output_file
  expansion
    search_phase           depth-first, best-first
    search_restriction     bounding-box, none
    granularity            #
*   minimize               time, space, wire-length
*   input_net_composition  two-point, multi-point
application
* type                     global, detailed
  alg_type                 gridded, graph
  sub_type                 gate-array, general
fabrication
* units                    metric, lambda
* pads                     yes, no
* connections              spanning, steiner tree
* technology               pcb, ic
*   xcapacity              #
*   ycapacity              #
*   zcapacity              #
*   pcb
*     number_of_layers     #
*     available_positioning fixed, random
*     layer
*       id                 #
*       expansion_direction x, y, z, 45
*       wire_width         #
*       wire_to_wire_space #
*       via_spacing        #
*       wire_to_via_space  #
```

Figure 5-1: User Specification Template (* required)

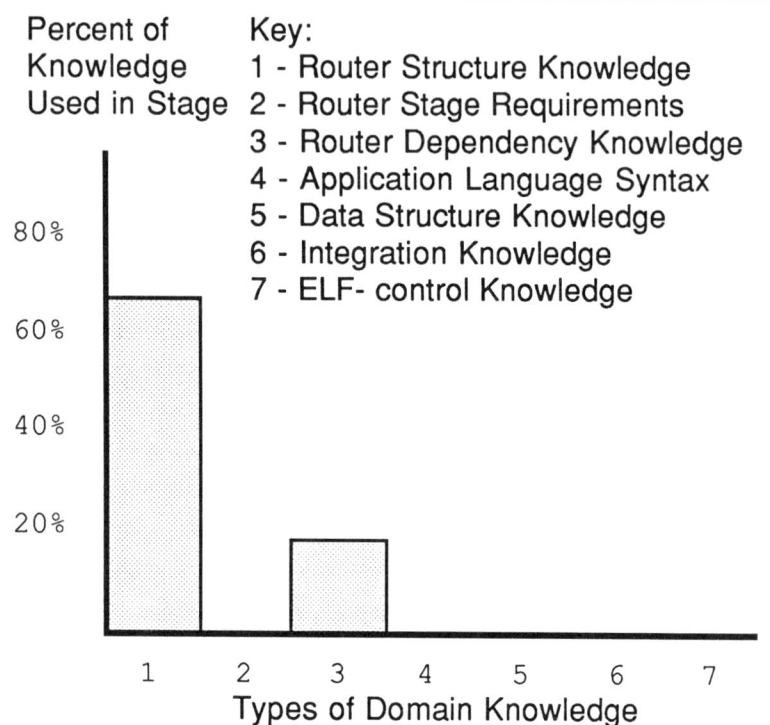

Figure 5-2: Types of Domain Knowledge used in the Input Stage

decision, many low-level decisions may be made (usually by applying domain knowledge) that are not individually accessible from the user-specification. This architectural detail limits user access to router decision subtasks, but greatly simplifies the typical user interaction. The merit of this type of user interface is an issue that is not addressed by this research.

The template is organized to reflect the three basic sets of technology constraints: algorithm, application, and fabrication constraints. Each entry is organized hierarchically first within its appropriate technology constraint set, then within its common subtasks. Examples of common subtasks are cost function determination and expansion search phase parameters. Specified entries not only affect the determination of unspecified entries in the same common subtasks, but also affect the determination of unspecified entries in the same technology constraint set and in *other* technology constraints sets. For example, in the following sample template (see Figure 5-3), the user specifies a 2 layer PCB, with diagonal (45°) wiring fabrication technology constraints.

```
application
    type                          detailed
fabrication
    technology                    pcb
    pcb
        number_of_layers          2
        layer
            id                    1
            expansion_direction   45
        done
        layer
            id                    2
            expansion_direction   45
        done
```

Figure 5-3: Sample Portion of a Specification Template

The search phase was not specified. The specification of diagonal wiring will initially imply a depth-first search scheme over a breadth-first scheme for the router search phase to avoid needless expansion of diagonal paths without first considering normal Manhattan paths. The depth-first search scheme avoids needless expansion, especially in detailed routing applications. The increased physical constraints imposed by diagonal wiring [3, 2] are used to infer a search phase solution which will use the diagonal connective paths only as a last resort. For example, Figure 5-4 shows two different wirings for the same source and target but different fabrication styles. Shaded areas shows those areas affected by wire spacing rules for the wiring path. The diagonal wiring affects a total of seven wiring cells, while the Manhattan-style rectilinear wiring affects a total of five wiring cells.

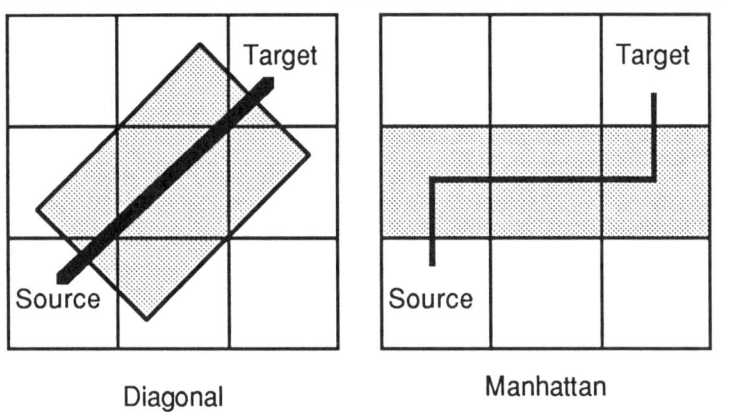

Figure 5-4: Increased physical constraints imposed by diagonal wiring

Inference requires a minimum specification set. These specifications are marked with an asterisk in Figure 5-1. Specifically, some routing expansion goals (i.e., minimize routing time, wire-length, space requirements, or some conjunction of these goals) must be specified, and in addition, some details of the target application and fabrication technology must be described.

While it is known to be important to make decisions as soon as enough background knowledge is known [1], the decisions made by the Input Stage inference engine are not irrevocable. To enable backtracking and future decision making within the Selection Stage, ELF generates multiple candidate options (e.g., both depth-first and best-first search for the search-phase task) for all unspecified tasks and subtasks, with the exception of fabrication options. Fabrication options are considered cast in silicon.

A user specification may generate conflicting design task and subtask options. This will cause the inference of more than one option for a particular task or subtask. Multiple options for a particular task or subtask are ranked by the strength of the inference. A highly ranked design option is regarded more favorably than a lower ranked design option by the selection mechanism within the Selection Stage. A router task or subtask option that was inferred directly from a user specification is ranked highest. Candidate task or subtask options that were derived from user specifications that were themselves inferred from other user specifications are next. Candidate task or subtask options that were inferred from other task or subtask options are last. Thus, the Input Stage provides a ranked ordering of candidates for each task and subtask to the Selection Stage.

5.2. Input Stage Rule Types

There are three types of rules in the Input Stage. The first type of rule deals with the reading and parsing of the specification template. These rules do not contain domain knowledge or program synthesis knowledge. Their function is to parse the user specification following the user-interface syntactic requirements as specified in Appendix I. An example is:

```
READ_in_parameter:
  Given have a command that requires a parameter THEN

  Read in parameter and place this information in
  working memory for later use in inference mechanism.
```

All user interface commands consist of either one or no parameters. Parameter options may be added without affecting the template parsing rules. These rules are of little interest to the overall design of ELF as they are purely a function of the user-interface template design.

The second type of rule uses router structure knowledge to run inferences on the incomplete user specification to generate candidates for any other unspecified user interface commands. An example of an inference rule is:

```
INPUT_INFER_net_conversion:
  Given INPUT_NET_COMPOSITION is multi-point nets AND
  Given SEARCH_PHASE is depth-first algorithm THEN

  Add a multi-point to two-point net conversion option
  and add conversion subtask to setup phase.
```

This rule is necessary because ELF's current depth-first search algorithms are based on a two-point search technique; multi-point nets must be decomposed in some optimal fashion into a sequence of two-point connections for routing.

The third type of rule is used to run inferences on the task and subtask options to generate other task and subtask options. There are two kinds of rules in this rule type. The first kind contains router dependency knowledge. This rule summarizes the results of option interactions as applied to tasks and subtasks. For example:

Automatic Programming Applied to VLSI CAD Software: A Case Study

```
INPUT_from_search:
  Given a spanning tree interconnect as subtask AND
  Given have multiple-point nets as subtask THEN

  Modify best-first expansion target check subtask
      to expand into target cell.
  Modify backtrack re-init subtask to include re-init
      subnets as starting points for future intra-net
      wirings.
```

This rule uses router dependency knowledge to identify the interaction between spanning tree interconnect and multiple-point input nets and generates the necessary subtask options to handle their interaction.

The second kind of subtask generation rule uses routing phase requirement knowledge to produce multiple candidate options for the same task or subtask. These rules interact with a predefined database of user selection possibilities. This database must be updated as ELF evolves. This is handled differently than the earlier parameter checking facility. A task or subtask can be forced to a particular candidate via a parameter in a template entry, but it can not be inferred or generated as a multiple candidate option until domain knowledge about its database repercussions are known. For example as new technologies are developed, new types of fabrication specifications (e.g., a three metal process) may be made available. It would not make sense to have the capability to infer this new fabrication specification before detailing the interactions between tasks and subtask options. But it does make sense to have the capability to force this specification, via the user interface template, to *test* out the current knowledge base as applied to the new fabrication specification. This rule generates all possible options for the search-phase task.

```
INPUT_multiple_option_search:
    Given no search selection in template AND
    Given no search selection is inferred THEN

    Generate all possible search options
            (depth-first, best-first, breadth-first)
```

5.3. Chapter Summary

This chapter discussed the basic organization of the Input Stage and its implementation within the ELF architecture. This stage uses inferences made from entries in the user-specification template to capture, at the highest level, interactions among the three sets of technology constraints; algorithm, application and fabrication. The Input Stage passes its (possible large, possible conflicting) set of candidates for all the required routing tasks on to the Selection Stage, described in the next chapter.

References

[1] D. Barstow, *Knowledge-Based Automatic Program Construction*, North Holland, 1979.

[2] H.H. Chen, "Routing L-Shaped Channels in Nonslicing-Structure Placement", *24th Design Automation Conference*, 1987, pp. 152-158.

[3] D.A. Edwards, "The Architecture of the Manchester Routing Engines", Tech. report, Univeristy of Manchester, England, September 1988.

Chapter 6
The Selection Stage

The Selection Stage chooses the appropriate data structures and router algorithms to implement the required routing tasks given the preliminary problem description produced by the Input Stage. Note, however, that one algorithm choice does not comprise an entire router. Indeed, there exists tens of possibilities for *each* of the many algorithms and data structures comprising a real router that matches any typical user specification. Hence, decision-making in this stage requires information on such topics as data structure capabilities, data structure efficiencies, and algorithm suitability. This stage integrates the *router structure knowledge, router phase requirement knowledge, data structure implementation knowledge* and *design interaction knowledge* sources to select and refine implementation candidates for both data structures and algorithms. In this context we define the terms selection and refinement as:

- SELECTION: The process of choosing between several candidate data structures or algorithms.
- REFINEMENT: The process of adding detail, or of modifying, the selected candidate data structure or algorithm. This is a result of applying some knowledge source.

It is essential to note that ELF does *not* make its selections for either algorithms or data structures from sets of implemented, executable modules. In other words, ELF does *not* select pieces of executable code, then somehow configure them and glue them together. Rather, we select broad *schemata* for algorithms and data structures and generate routers based upon the schemata selected. We discuss issues relating to the selection and refinement processes in the following sections.

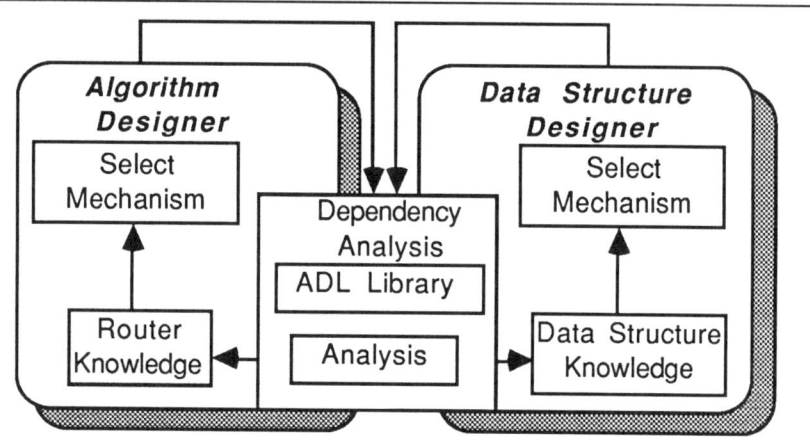

Figure 6-1: Selection Mechanism

In the ELF architecture, we explicitly acknowledge the interdependence of algorithm choices and data structure choices [2]; indeed, this is the crucial problem to be handled during the selection process. Figure 6-1 illustrates the selection process. The Input stage generates a set of candidate alternatives for each design task. These design tasks and subtasks further map to sets of data structures or algorithms. Such an ordered set of alternatives for a particular data structure, or for a particular algorithm, is termed a *candidate set*. Selection of the appropriate data structures and algorithms is implemented by four modules. They are: the *Selection Control Module*, the *Dependency Analysis Module*, the *Data Structure Designer Module*, and the *Algorithm Designer Module*.

The Selection Control Module controls the process. For each alternative in these candidate sets, data structure interdependencies and timing estimates are formed by the Dependency Analysis Module. This information is combined with *router domain knowledge* within the Algorithm Designer Module to select the more promising algorithm alternatives, thereby pruning unpromising algorithm alternatives from the appropriate candidate sets. Similarly, the Data Structure Designer Module uses this interdependency and timing information, combined with *program synthesis knowledge* of data structure capabilities, to select data structure alternatives. The Algorithm and Data Structure Designers are invoked iteratively by the Selection Control Module, i.e., information on the most recent algorithm selections is fed back to the Data Structure Designer to assist its own decisions, then information about subsequent data structure selections is fed back to the Algorithm Designer, and so on. This iteration continues until final selections on all candidate sets are completed. Iteration is controlled via the Selection Control Module.

Figure 6-2 shows the types of domain knowledge utilized in the Selection Stage. This chapter discusses some Selection operational issues along with knowledge source application and integration. The following sections detail the operation of the Selection Stage along with illustrative examples of the domain knowledge within each of its four modules.

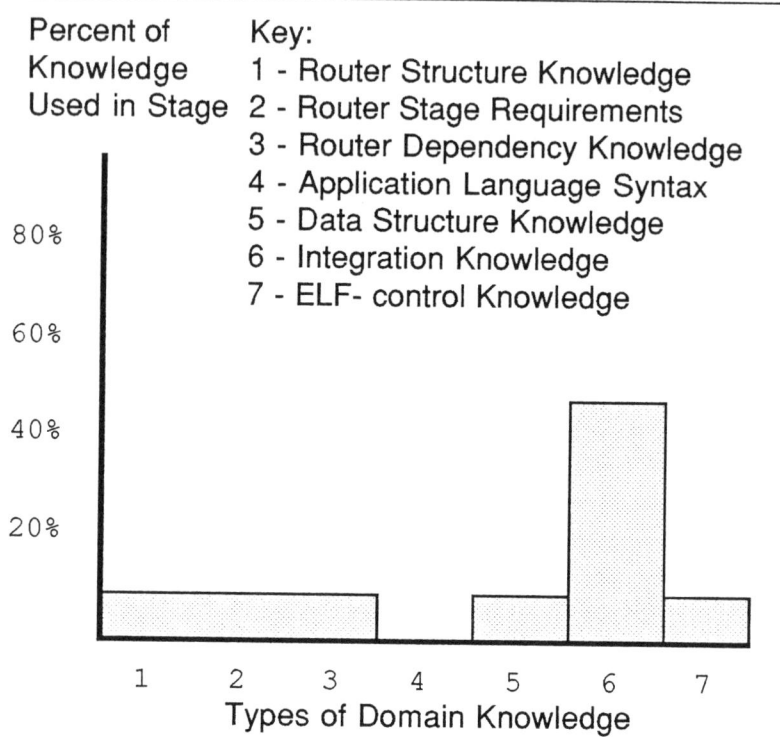

Figure 6-2: Types of Domain Knowledge in the Selection Stage

6.1. Selection Control Module

The Selection Control Module regulates the interaction between the data structure and algorithm selection processes [2]. To ensure that constraints generated during one selection process are considered in the other selection process, we restrict each Designer to a single decision per iteration before control is switched to the alternate Designer Module (see Figure 6-3). Switching control to the alternate Designer Module allows the ELF architecture to react to a single selection and refinement decision for an algorithm or a data structure style.

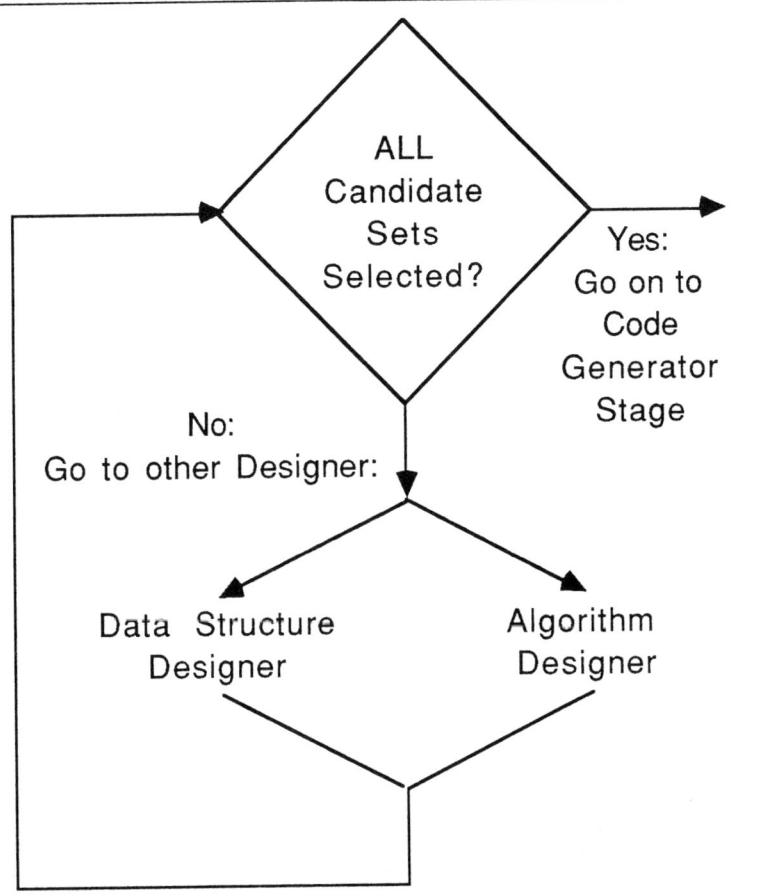

Figure 6-3: Selection Control Module Process

In addition, the Selection Control Module recognizes when the router design tasks have been completed and are therefore ready to go on to the Code Generation Stage. Completion does not signify full specification, as the Code Generator Stage still requires *router domain knowledge* to transform the selected data structures and algorithms to their executable format. Design tasks are considered to be complete when all refinements of the task design, whether they be data structure refinement or algorithm modification refinement, have been attempted. Figure 6-4 shows the type of domain knowledge, as described in Section 4.5, utilized in this Module.

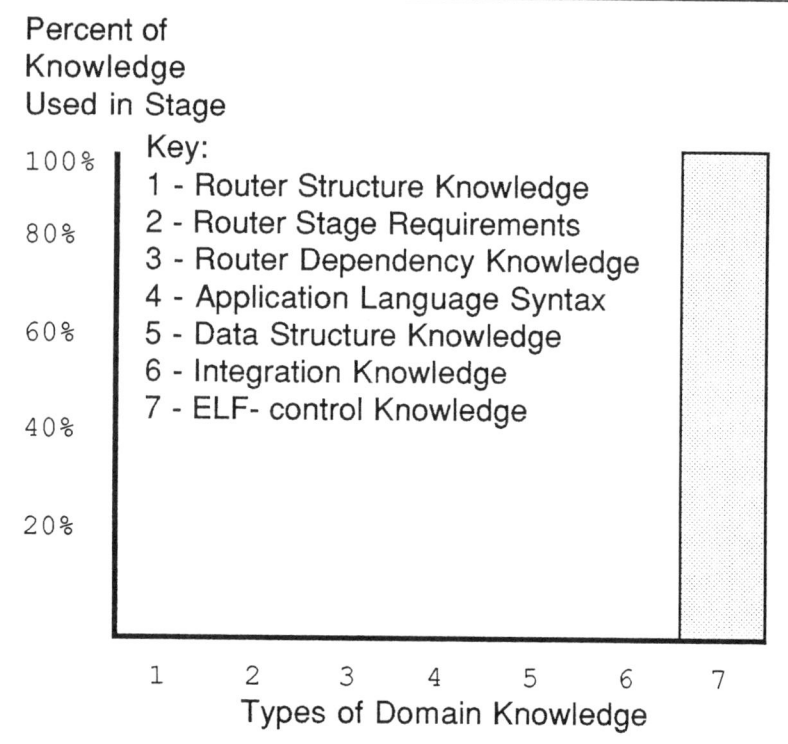

Figure 6-4: Knowledge in the Selection Control Module

There is one basic type of rule in the Selection Control Module which controls the interaction of the Algorithm and Data Structure Designers. This rule type recognizes the interdependency of algorithm and data structures and resolves critical dependencies by switching appropriately between the Algorithm

Designer Module and the Data Structure Designer Module. For example:

```
DESIGNER_CONTROL_iterate:
      Given a refinement by either Designer AND
      Given more refinement is required THEN

      Cost undecided alternatives by starting
           up Dependency Analysis Module.
      Start up the other Designer Module.
```

This rule notes that some candidate set refinement was performed by either the Algorithm Designer or Data Structure Designer, but that not all refinement has been completed, i.e., not all routing sub-tasks have been completely specified. Therefore another round of refinement is required. Results from the previous refinement are first combined with the ADL representations to derive new costs for all candidates in candidate sets where a final decision has not been made. Currently, the Data Structure Designer Module has precedence over the Algorithm Designer within each round of refinement. The Data Structure Designer Module is allowed to make a refinement first. The other Designer is then given a chance to make a refinement based on the any new actions taken by the current Designer.

If no candidate set refinement is performed in a particular iteration, then the Selection Control Module makes an arbitrary decision by choosing the highest-ranked candidate from an arbitrary candidate set, and then attempts to continue with the refinement process.

6.2. The Dependency Analysis Module

This Module is a slave to the two Designer modules. No search is performed by this Module, it is simply told what ADL algorithm representation to analyze. The Module has two goals. The first goal is to compute the basic "cost" of the ADL algorithm representations, including costs of all data structures. The second goal is to provide information on set composition and set interdependencies. Recall that in the ADL language, there is no explicit data structure implementation. Each data structure is accessed and defined as sets of fields. Completion of these goals requires the Dependency Analysis Module to analyze the ADL representation of each algorithm alternative. Its output is the cost and interdependency information.

The first type of rule here computes the costs associated with algorithm and data structure representations. The "cost" of an ADL algorithm representation is represented by two metrics: the best anticipated time estimate, and the worst anticipated time estimate. These metrics are summarized over the entire algorithm representation using the data structure selections as currently refined. The time estimate is derived from information about each algorithm's usage, e.g., data structure iteration or access. These metrics require two sources of information, namely, which algorithms to analyze (Algorithm Designer Module), and which data structures to use (Data Structure Designer Module). Computing the two metrics requires *application language syntactic*, *data structure implementation*, and *design interaction knowledge*. Figure 6-5 shows the type of domain knowledge, as described in Section 4.5, utilized in this Module.

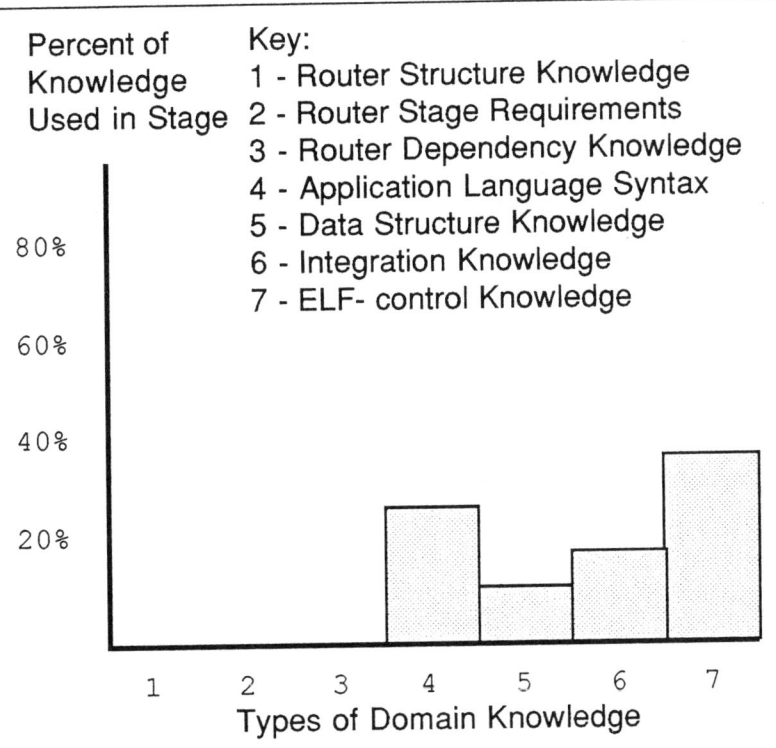

Figure 6-5: Knowledge in the Dependency Analysis Module

The cost associated with an algorithm is critical to the pruning process. The Dependency Analysis Module refines this metric as data structures are refined, as the refinement process itself produces a more concrete base to establish algorithm costs. Examples of this first type of rule in the Dependency Analysis Module are:

```
DEPENDENCY_algorithm_cost:
    Given an ADL algorithm representation AND
    Given the algorithm consists of a major block AND
    Given the major block is a LOOP statement THEN

    The ADL algorithm representation cost metric is
        equal to the timing costs of accessing
        the set associated with the LOOP statement.

DEPENDENCY_modify_algorithm_cost:
    Given a refinement of a data structure AND
    Given an ADL algorithm representation cost metric
        that is dependent of that data structure THEN

    Update the cost for the ADL algorithm
        representation to reflect the refinement.
        (This is functionally equivalent to a recost)
```

The first rule example shows the dependency of the timing cost estimate for an ADL algorithm representation on the data structure chosen to implement the major LOOP statement. The second rule example reflects the need to maintain the validity of cost metrics as algorithms and data structures are selected within the Designer Modules.

The "cost" of possible ELF implementations for the data structure sets used within the ADL representations is represented by three metrics. These metrics are: the best anticipated space utilization, the worst anticipated space utilization and the access speed utilization. The space metrics are in memory words required, and the speed utilization is the memory access time required to identify an element in the set. We compute the first two metrics using *data structure implementation* and *design interaction knowledge*. For example, the space consumption of an array is evident just by examining its bounds, while the space consumption of a list must be estimated by *router domain knowledge* of that particular set's sparsity characteristic. We compute the final metric, access speed utilization, by mapping the ADL representation to *router domain knowledge* of its intended use to arrive at an approximation of the impact of a particular data structure on the algorithm timing requirements. This metric maps the algorithm cost metrics to the data structure cost metrics. For example, a best-first algorithm significantly stresses the data structure representation of the routing area, the grid. The impact of this data structure representation on the best-first algorithm timing is given by the access speed utilization metric. Computing this metric requires *router domain knowledge*, *design interaction knowledge*, and *data structure implementation knowledge*.

The second type of rule computes the root "cost" of a possible ELF implementation for the data structure sets used within an ADL algorithm representation. The Dependency Analysis Module refines this metric as data structures are refined. Generic *program synthesis knowledge* about data structure implementations underlies the computation of the cost function. An example of a cost computation rule is:

```
COSTING: structure_iterated_array_time:
    Given accumulating costs for a set AND
    Given array is available as the root type AND
    The set is ordered by X with range <x> AND
    The set is not sparse THEN

    The time estimate for the array is 1.

COSTING: structure_iterated_list_time:
    Given accumulating costs for a set AND
    Given list is available as the root type AND
    The set is ordered by X with range <x> AND
    The set is not sparse THEN

    The time estimate for the list is <x>.
```

This rule computes the relative costs for an array and a list, given domain knowledge of this particular set's probable application. This rule estimates the time required to access a set for the available implementations array and list. *Router domain knowledge* that this set is organized via the value <x> supplies the necessary access timing information. This rule demonstrates the interdependence of program synthesis and domain knowledge in data structure costing. While *program synthesis* knowledge drives the cost computations, such knowledge alone can not fully derive an accurate cost without domain knowledge of the expected set's application.

Finally, the Dependency Analysis Module provides information about set composition and set interrelationships. This information supplements the data provided by the ADL interrelationship segment described earlier in Section 4.2. We allow set interrelationships to be one of:

- **Intra-set assignment**: Assignment between two fields of an abstract set. For example, the variables x and y must access the same values, and they are in the same abstract set:

    x OF set = y OF set

- **Inter-set assignment**: Assignment between two fields of different sets. For example, the variables x and y must access the same values, yet they are in different abstract sets:

    x OF set = y OF different_set

- **Intra-set iteration**: Iteration through a set via a field of the same set. For example, the variable x accesses each element of the abstract set:

    LOOP x OF set IN set

- **Inter-set iteration**: Iteration through a set via a field of a different set. For example, the variable x accesses each element of a different abstract set:

    LOOP x OF set IN different_set

All of these interrelationships denote some type of implementation data structure dependency between the abstract sets. This information may be combined with *router domain knowledge* to assist the Data Structure Designer Module in both selection and refinement of set-internal variables. For example, the ADL statement

    Assign path_not_found to check_target

will cause an inter-set assignment dependency between the two data structures. Further analysis shows that check_target is a function returning a value. Selection of the data structure, actually a variable, path_not_found, is dependent on the type of value returned by check_target. This conclusion represents an cumulative analysis of several ADL representations.

The third type of rule notes all data interrelationships which might later be used in the Data Structure Designer Module to prune selection. Two examples of this type of rule are:

```
DEPENDENCY_rhs_to_lhs:
    Given a left-hand of a statement AND
    Given a right-hand of a statement THEN

    The left-hand-side type must be a
      subset of the right-hand-side.

DEPENDENCY_loop_iteration:
    Given a loop statement AND
    Given a loop iterator AND
    Given a set THEN

    The loop iterator must be type-compatible
      to iterate through the set.
```

These rules define the loop iteration rule. The first rule states the basic data structure type equivalence for any particular statement. The second rule requires that any structure variable that iterates through some data structure must be type compatible.

6.3. The Data Structure Designer Module

The Data Structure Designer Module successively selects and then refines the initial candidate sets of abstract data structure representations for each data structure required by any candidate algorithm schema. For each essential data structure in the routing task, such as the routing grid, the wavefront of expanded cells, etc., a specific implementation must be chosen based on the needs of particular (and dependent) algorithm choices. For example, a backtrace data structure implementation for a depth-first maze router has different time/space characteristics than one for a breadth-first maze router. The algorithm selected for wavefront expansion affects the implementation chosen for the backtrace data structure. Again, this illustrates the interdependencies between data structure and algorithm selection.

Data Structure selection is built upon two information sources. One is the ELF knowledge base consisting of *router domain, program synthesis,* and *design interaction knowledge.* The second source is the interdependency information gleaned from the Dependency Analysis Module. The Data Structure Designer Module must:
1. Organize the interdependency information gleaned from the Dependency Analysis Module.
2. Integrate the ELF knowledge base with the interdependency information.
3. Select a candidate from candidate set.
4. Refine a candidate to actual implementation.

This section reviews these four issues and gives illustrative rules detailing the use of domain knowledge within the data structure selection process. Figure 6-6 shows the types of domain knowledge used in the Data Structure Designer Module.

6.3.1. Representation of Data Structure Interdependency Information

Data structure interdependency information aids the data structure process by capturing set relationships found in the ADL algorithm schema representations. These set relationships are exploited to quickly transmit the effects of domain knowledge-driven data structure selection operations to other data structures. A representation of the interdependency information must have the following fundamental properties:
- Provide an abstract representation of data structures.

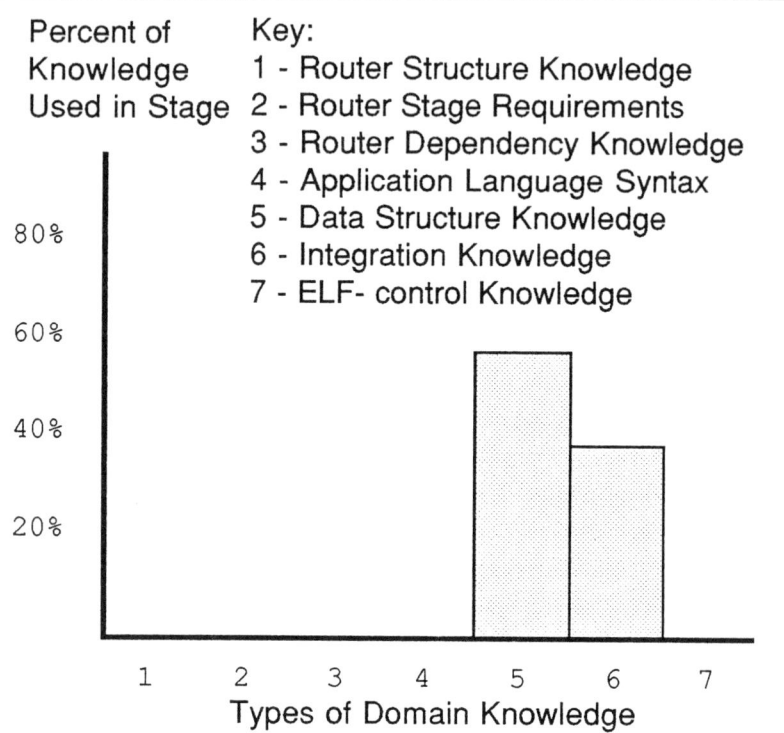

Figure 6-6: Domain Knowledge in the Data Structure Designer Module

- Capture interdependencies between abstract data structure selection candidates.
- Identify the type and strength of the interdependency.
- Reflect the effect of data structure interdependencies on data structure composition.

The strength of a data structure interdependency is a measure of the mutual interdependency. For example, in a database [1], the form of a piece of retrieved data depends mainly on the form of the query that generated it. Similarly, two data structures may be strongly related if one data structure selection dictates the other selection. A weak relationship is present if components of the data structures reference each other, but do not coerce the other data structure to any particular form.

ELF builds a data structure interdependency representation model shown in Figure 6-7. It is graph-based with a node for each data structure, and an arc for each interdependency. Each arc is labeled by the relationship interdependency strength, either *strong* or *weak*. Within each node is a current abstract representation of the data structure. This abstract representation reflects the current interdependencies affect that particular data structure's composition.

The data structure interdependency graph representation serves two purposes. First, it provides a method of organizing interdependency information. Second, it provides a framework for the hierarchical data structure composition produced by data structure selection. Data structure representation is detailed in Section 6.3.2. Each node of the interdependency graph represents a single abstract data structure and contains the current hierarchical data structure representation for each candidate. Data structure selection operates and modifies the interconnectivity graph to reflect domain knowledge-driven selections.

6.3.2. Data Structure Representation During Selection

One selection issue is how candidate data structures are represented within each node of the interdependency graph. The representation must readily accommodate successive selection and refinement operations. When ELF selects a data structure representation within a node of this graph, it builds these representations top-down. Consider the synthesis of the wavefront structure shown in Figure 6-8. The first decision is whether the *root structure* is an array

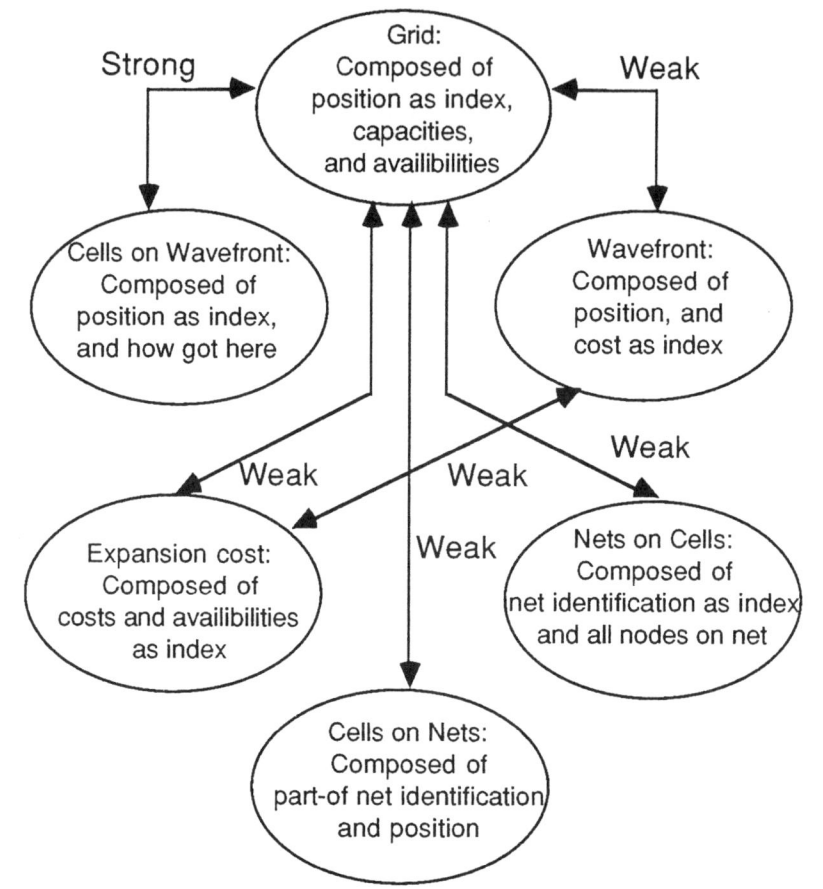

Figure 6-7: Basic ELF Data Structure Interdependency Graph

or a list; the decision depends in large part on the expansion search scheme, and how cost values will appear in cells. Any particular element of a data structure can either appear as an explicitly defined field within the data structure, or it can appear implicitly as the ordering (i.e., dimensioning) value used to define a particular element within the data structure. (Recall how the maze router algorithm requires some cost ordering for cells being expanded.) In this

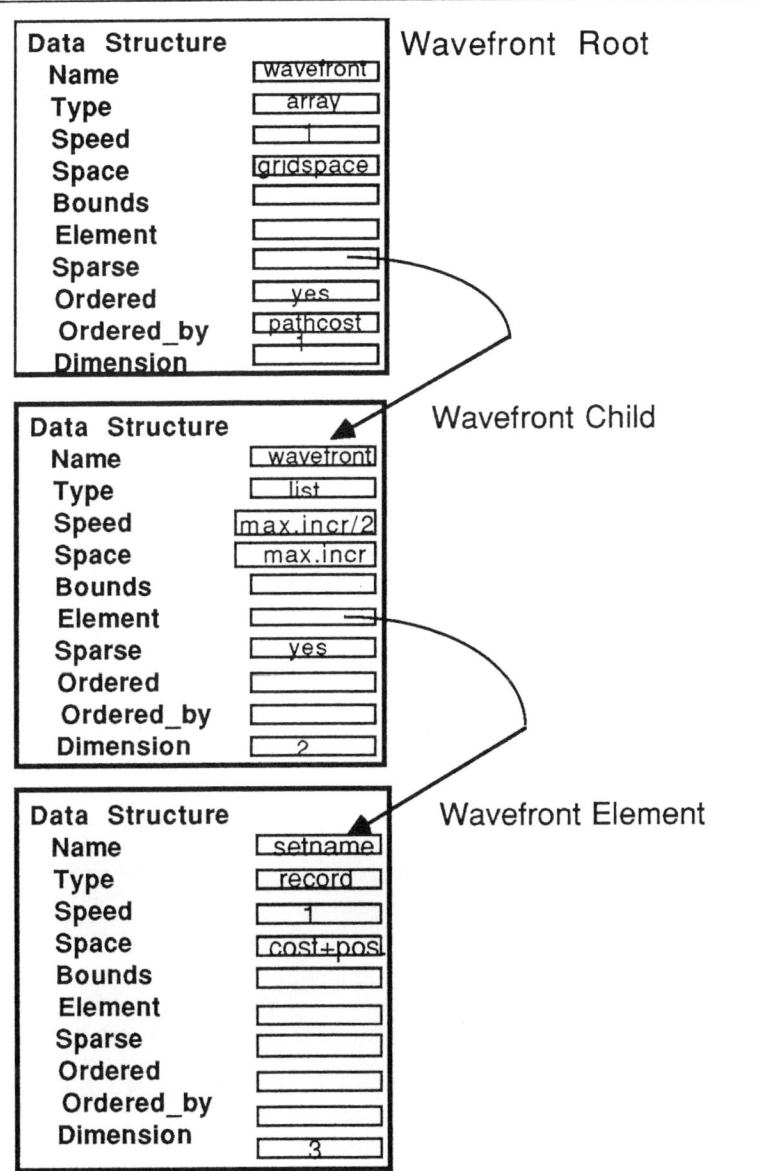

Figure 6-8: Data Structure Building: the Wavefront Data Structure

example, ELF concludes (using *router domain knowledge*) that the wavefront structure will probably be indexed by (or is dimensioned by) a path-cost value (an aggregate value for the cheapest partial routing path found so far). Hence, each element of this root structure is another *child structure* that holds all cells reachable with a particular path cost. This root structure will also likely be ordered so that the least expensive cell to expand can be found quickly. In addition, depending on the cell expansion scheme, the child structures may need a special ordering as well. All these constraints must be considered in choosing reasonable implementations for the root and child structures.

Selection allows three basic data structure styles, namely, list, array and record. Selection refinement yields 9 possible data structure styles after one refinement step and yields 20 possible styles after two refinement steps. (Several possible styles are not reachable at each refinement step. For example, a one step refinement to a record cannot then produce another refinement step to a record of a record.) The refinement process uses domain knowledge to control the overall process, and requires only simple cost computations. Figure 6-9 details the available data structure options for an arbitrary data structure. Each data structure has a root type. ELF allows the root type to be either an array, record, or a list. Because the set of root types is small, each data structure can be easily analyzed using the data structure cost metrics produced by the Dependency Analysis Module. These cost metrics along with domain knowledge produce a ranked ordering of the three possible root types. Specifically, the best and worst space utilization and the access utilization metrics are evaluated against the user directive (either to minimize routing time or to minimize space utilization) given in the input template. The metric that best suits the user directive is the preferred choice and is ranked the highest. Normally, the highest ranked root type is chosen for refinement. However, there are cases where domain knowledge produces contradictory rankings, e.g., when both time and space are to be minimized. In this case, both top rankings are considered and the root decision is delayed until the final refinement process.

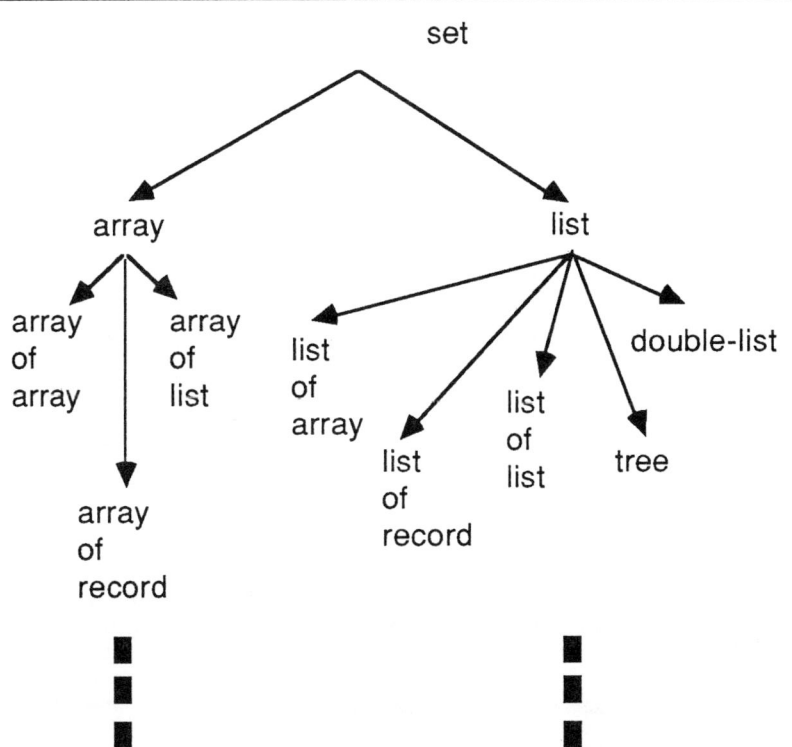

Figure 6-9: Basic Data Structure Selection Possibility Tree

The root type is refined using the data structure cost metrics as well as domain knowledge of the data structure size and hierarchy. Refinement continues until the data structure is completely specified for its intended task, i.e., each dimension of the structure has a selected implementation.

Building complex data structures such as these requires more information than just the relevant high-level algorithm descriptions found in ADL, and the basic characteristics of elementary data structures. Together, generic *program synthesis knowledge, router domain knowledge*, and *design interaction knowledge* guide the selection and refinement of these data structures. This section reviewed the data structure selection process and the role of domain knowledge in the selection and refinement of those data structures required to complete the routing task.

6.3.3. Data Structure Selection Operation

Selection of data structures is a two tier process. First, a data structure candidate must be preferred over the others. This is called *SELECTION*. Then, the candidate data structure and its elements are refined to the final representation. This is called *REFINEMENT*. *Router domain knowledge* and *Data structure implementation knowledge* aid both selection and refinement. This section outlines the seven steps required to produce first the selection, and then the refinement of data structures, what domain knowledge is required within each of the seven steps, how this domain knowledge is integrated with the interdependency information, and how data structures selections and refinements are represented within each node of the interdependency graph. These steps are shown in Figure 6-10.

The first three steps are only performed upon initialization of the Data Structure Designer Module. Successive steps are iterated throughout the

1. Build basic starting interdependency graph (See Figure 6-7).
2. Delete those nodes with no connectivity.
3. Put the node with the highest degree of interconnectivity into the Active subset.
4. Either:
    a. Select highest ranked candidate implementation for an unselected node in the Active subset or
    b. Refine candidate implementation for the incomplete node with the largest degree of interconnectivity.
5. Rerank the set of candidate implementations for each node that is dependent upon the nodes just modified in 4.
6. Place all interdependent nodes in Active subset.
7. Repeat steps 4,5,6 until all data structure nodes are selected and refined.

Figure 6-10: Data Structure Selection Decisions

selection process. The first step builds the interdependency graph using the information provided by the Dependency Analysis Module. The set of ADL candidate algorithm representations are individually parsed, while looking for keywords. Keywords denoting relationships are ASSIGN, TEST, LOOP, MAPPED_BY, and PARTOF. The first three are action keywords as they cause data structure manipulations, either directly or indirectly. Such data structure manipulations are indicative of strong dependencies. A LOOP statement causes a strong dependency between the data structure set and the index. The ASSIGN statement causes a strong interdependency between the left-hand-side and the right-hand-side data structures. The final two are element-relating keywords that may exist within the ADL syntax of the action keywords. For example, the partial ADL statement

ASSIGN wavefront MAPPED_BY expandsite TO costfunction

causes both a strong dependency between the `wavefront` and `cost_function` data structures, and a weak dependency between the `wavefront` and `expansion_site` data structures. The `wavefront` data structure references the `expansion_site` data structure, but it does not produce the data structure.

An example of a rule building the interdependency graph is:

```
DS_SELECTION_dependent_ds:
    Given an ASSIGN keyword AND
    Given a left-hand-side data structure (lhs) AND
    Given a right-hand-side data structure (rhs) THEN

    Make a strong dependency between lhs and rhs.
```

This rule captures a type of strong assignment interdependency between the left-hand-side and right-hand-side data structures.

Graph nodes that do not have any arcs to other data structures are deleted. These correspond to nodes that were defined in the ADL interrelationship segment, but were never used in this subset of ADL representations. Common examples are the pad data structure representation and the wiring track data structure representation. These data structures may not be used in the desired application and may be removed from the selection process. Deletion of the representative graph node removes a data structure from the selection process.

ELF uses a *greedy*-type algorithm in selecting the selection *seed* node to start the overall selection process. The first node selected is always the `grid` data structure. The data structure interdependency graph does not drastically change as input parameters are modified. Some arcs disappear and change strengths, but the node with the highest interconnective degree is generally the `grid` data structure. The result is that the `grid` implementation selection usually drives the other data structure selections. The `grid` data structure represents the mapping of the abstract notion of expandable cells to the physical routing domain. The `grid` implementation selection is a direct result of technology restrictions and is most closely tied to technology, requiring the flexibility to change with technology, e.g., the number of layers, or changes to wiring constraints. Thus, selecting the `grid` as a seed to the selection process ensures that the selection will efficiently adapt to changes in technology without excessive selection control iterations.

A **select** or **refine** decision is made upon each successive entry into the Data Structure Designer Module. Each decision is based upon the integration of the information presented in the interdependency graph with either *router domain knowledge*, *program synthesis knowledge* or *design interaction knowledge*. The result of either decision is reflected in the current data structure representation within the applicable node of the data structure interdependency graph. The following sections exemplify these knowledge sources within the selection process.

6.3.3.1. Router Domain Knowledge

Router Domain Knowledge encompasses the most basic requirements for data structures supporting practical routers. This knowledge maps router concepts to generic program synthesis concepts. An example is the elementary domain knowledge that for a detailed maze router there must be data structures for the routing grid, for the expanding wavefront, and so forth. For example, in the selection of the wavefront data structure, knowledge of the set's ordering, e.g., the likely ordering of cells, is this type of knowledge.

This rule type refines the root type of a given data structure using *router domain knowledge* to transform it to a more concrete form. Refinement can take two forms. First, the structure type can be made more explicit, e.g., an array becomes an array of lists. Second, the internal structure can be recomposed so as to invert fields within the data structure implementation. We treat the idea of field inversion in more detail in section 6.3.3.5. The following rule uses Dependency Analysis Module metrics in combination with user directives to refine data structure selections.

```
DS_SELECTION_layers_influence:
  Given the number of routing layers <= 2 AND
  Given a gridded router AND
  Given a low access speed utilization AND
  Given a sparse dimensional characteristic
     for a particular data structure AND
  Given the particular data structure is
     selected ARRAY THEN

  Refine the selection to an ARRAY OF LISTS.
```

This rule refined an ARRAY selection to an ARRAY OF LISTS based on a combination of knowledge sources, specifically:
- Router domain knowledge that the data structure was sparse.
- Access speed utilization numbers derived by the Dependency Analysis Module.

6.3.3.2. Program Synthesis Knowledge

Program synthesis knowledge evaluates the costs associated with the data structure implementations to rerank candidates. This knowledge maps the known data structure implementations to functions detailing their approximate single access times, iteration times and space requirements. Examples include: generic *program synthesis knowledge* that an array can be immediately accessed knowing the indices, that the time to iterate over a data structure is the product of its specified dimension sizes, and that the space requirement of a data structure is the product of its specified dimension sizes and the element size. The dimension size information is often specified by router domain knowledge.

```
DS_SELECTION_first_dimension_array:
    Given set S AND
    Given set is ordered by X with value <x> AND
    Given candidate array option AND
    Given variable V that VARY_BY set S

    Select integer for V.
```

This rule looks at the information provided by the data structure interrelationship segment in the ADL candidate representations, in this case variable V ranges over the set S, to select the appropriate data structure for V, namely that V should just be an integer.

6.3.3.3. Design Interaction Knowledge

Design interaction knowledge concerns the effects of design tradeoffs as seen in the interplay between data structure implementations and algorithm selections. As an example, consider the impact of allowing 45° wiring in addition to rectilinear wiring. The router must examine cells that are diagonally adjacent in the routing grid during the expansion phase, since paths can now make 45° bends. This has an obvious effect on the expansion algorithms, and on all data structures having to do with the wavefront. The addition of diagonal adjacency increases the information needed within each element of the data structure that represents the evolving routes, and changes the space and time characteristics of this data structure. This may result in the selection of a more space-conserving implementation.

This rule type guides candidate set decisions based on the metrics derived by the Dependency Analysis Module and user input specifications. For example, the following rule combines the user directive to minimize space with domain knowledge of probable sparsity--that the chosen search will lead to a sparse data structure structure--to select a linked-list as the root type for a particular data structure. For example:

```
DS_SELECTION_wavefront_as_list:
    Given best-first expansion AND
    Given minimizing space consumption AND
    Given the set wavefront may be sparse THEN

    Select list as the wavefront root type.
```

Note the interdependence between algorithm and data structure candidates implied by this rule. This rule uses Algorithm Designer Module alternatives along with domain knowledge of set sparsity to suggest a root data structure implementation for the set wavefront. Data structure alternatives can and do affect algorithm selection, given some optimization goal, and the efficiency characteristics of the data structure implementation alternatives. Similarly, algorithm alternatives, combined with knowledge of how algorithms access data structures, affect data structure selection.

6.3.3.4. ELF-control Knowledge

ELF-control knowledge ranks implementation candidates within a candidate set. Recall the candidate set is represented within each node of the data structure interdependency graph. Reranking can occur whenever new tasks or subtasks are created by the Algorithm Designer Module, or whenever the highest ranking candidate was unsuitable for the routing design task. For example:

```
DS_SELECTION_reorder_costs:
  Given need to rerank a candidate set THEN

  Recompute candidate data structure implementation
     costs.
  Rerank using user-specified ranking method
     (speed or space utilization).

DS_SELECTION_rank_candidates:
  Given reranked candidate set AND
  Given user preference to maximize routing speed AND
  Given computed speed costs for each candidate AND
  Given previous highest ranked candidate AND
  Given highest ranked candidate that is not the
     previous highest ranked candidate THEN

  Rank the candidates in order of costs, with the
  exception of the previous highest ranked candidate
  is ranked the lowest.
```

The first rule causes a recomputation of the data structure candidates within the desired candidate set. The actual recomputation is handled within the Dependency Analysis Module. The second rule reorders the candidates based on the user-specified ranking method of maximizing router speed. The previous highest ranked candidate, if any, is not allowed to remain the highest rank.

6.3.3.5. Operations on the Interdependency Graph

Finally, we note that each of these knowledge sources may be combined with the information in the interdependency graph to perform either selection or refinement. For example, *router domain knowledge* causing the selection of an array implementation for the wavefront may be combined with the structure of the interdependency graph to cause a compatible selection for each data structure strongly dependent upon the wavefront data structure. In cases where the interdependency graph causes incompatible selections for a data structure, program synthesis information about the costs associated with each of the selections may force a selection and then generate any necessary translation routines. Weak interdependencies can influence the composition of data structure refinement. For example, an array of linked-lists may be the selection for the wavefront cost data structure. Data structures weakly interdependent on the wavefront cost data structure must be able to reference each element of the wavefront cost data structure. Unique identification of each element requires elements of those weakly interdependent data structures first to have an integer value to access the array, then a pointer value to access the linked-list.

The **select** option has precedence over the **refine** option. This precedence rule encourages the proliferation of interdependency information at the highest level of data structure design, namely, the selection of the root candidate.

An important byproduct of the refinement process is field inversion. For efficiency reasons, it may be advantageous to invert fields within an element of the set as the set is composed. For example, in Figure 6-11, the wavefront is defined as a set of cells with each cell having a current cost field. *Router*

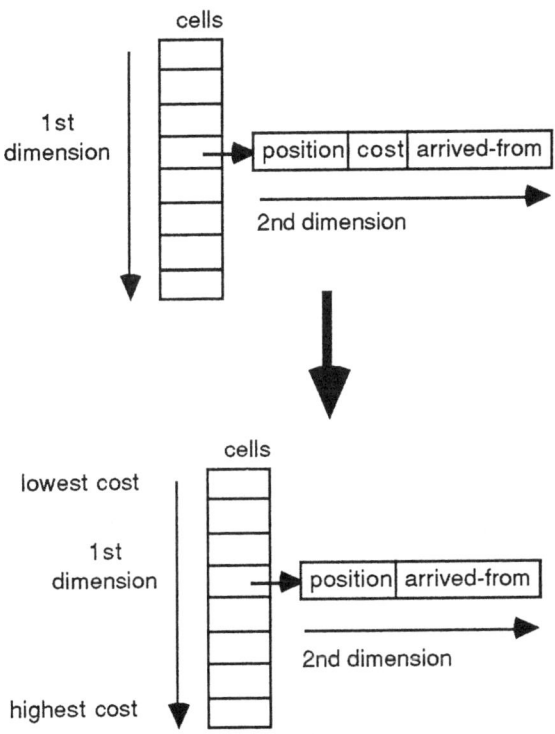

Figure 6-11: Field Inversion

domain knowledge refines this definition to an *ordered* set of cells, and recognizes the current cost field as the ordering element for each other field within the set. This field is then inverted so that it now actually defines the first layer--dimension--of this hierarchically-built data structure. For example:

```
DS_SELECTION_invert_field:
  Given an element of a data structure has a
      weak interdependency with the same
      data structure AND
  Given router domain knowledge that data structure
      has the same weak interdependency with every
      other field in the same set THEN

  Try a field inversion:
      Remove the element and place it within
      the same dimensioning range as the
      weak interdependency.  This location is
      router domain knowledge dependent.
```

This rule tries a field inversion when an element within a data structure has a weak interdependency with every other field in the same set. In this case, the element defines a dimension, or hierarchical layer. Field inversion removes the element from its current position in the data structure's hierarchical representation and, using *router domain knowledge*, moves it up to an organizational function within the hierarchy. As an example (see Figure 6-11), the set wavefront consists of the following elements: a cell position number, the cell wavefront cost, where the wavefront arrived from, and the cost of a change in direction. The set may be indexed either by the cell position number or the cell wavefront cost. The cell position number may have a weak interdependency with the wavefront data structure. The second rule will cause the cell position number to be removed as an element of the wavefront data structure and placed as an implicit indexing element to the data structure.

Another important byproduct of the refinement process is the creation of any needed *internal access* fields. Internal access fields are variables that are required to implement the data structure, but were never directly referenced by the ADL description. For example, in the representation of a sparse array as a linked list, one of the fields of the list element must be the column index. This variable must be *created* by the refinement process and correctly accessed by the code generator. Both trees and double-linked lists can be *refined* from a list root type by the creation of appropriate internal element specifier variables that can result only from application of domain knowledge. However, neither selection is available as a top level root choice. This rule type creates any new data structures that are required to justify prior selections and refinements. Different alternatives may require the creation of internal support variables. For example:

```
DS_SELECTION_follow_list_of_array:
  Given a set is list of array THEN

  Create head of list pointer.
  Create within-list pointer as part of set element.
```

In this rule, some set implementation was refined to be a list of arrays. Two new internal variables were created. The first is a pointer to the head of the list. The second is a pointer to the next element in the list, to appear within each set element. These internal variables will be utilized by the Code Generator Module whenever an element of this set is referenced and accessed in ADL.

6.4. The Algorithm Designer Module

The Algorithm Designer Module is the companion Designer to the Data Structure Designer Module and is used to select and refine algorithm candidate sets. Similar to the Data Structure Designer, this Module employs a two tier process that first selects the initial candidate algorithms and then, using domain knowledge, refines the candidate set of algorithm schemas for each sub-task in the overall router. Algorithm schemas are chosen by some measure, i.e., domain knowledge, of suitability and applicability. Selection is initiated by the *Input Stage*, and continues in the Algorithm Designer Module. Example tasks include: application data management, router expansion, and router backtrace. We combine *router domain knowledge* with user specifications and other Algorithm Designer decisions to propose new algorithm candidates. For example, selecting a depth-first type of expansion search places restrictions on input netlist composition, i.e., they cannot be multi-point nets. Therefore selecting a depth-first expansion search algorithm will cause a new candidate algorithm to be added to the candidate set for the router input sub-task: a conversion algorithm to transform multi-point nets into two-point nets. The second decision process *refines* the candidate set using several sources of information to prune the algorithm implementation candidates. Information sources used include user specifications, domain knowledge, previous data structure refinements, and algorithm cost computations. The Dependency Analysis Module provides the algorithm cost computations necessary to rank each member of the candidate set.

6.4.1. Algorithm Representation

One central issue is how candidate algorithms are represented within the Algorithm Designer Module. The representation must readily accommodate successive selection and refinement operations. Algorithms are represented within the ELF system using the ADL abstract representation language. Within the Algorithm Designer Module, algorithms are represented as shown in Figure 6-12.

Each algorithm representation template consists of a path name to the ADL

Algorithm Representation:

```
ADL pathname                    [        ]
Cost metrics:
   Timing best                  [        ]
   Timing worst                 [        ]
   Space best                   [        ]
   Space worst                  [        ]
Task name                       [        ]
Task ranking                    [        ]
Knowledge applied               [        ]
```

Figure 6-12: Algorithm Designer Module Internal Algorithm Representation

representation, its associated cost metrics, its current task ranking, and its task or subtask name. In addition, the template maintains a list of all applications of domain knowledge, whether *router domain*, or *program synthesis knowledge*, or *design interaction knowledge*. In this way, infinite algorithm reordering due to domain knowledge application, then reapplication, is avoided. The Algorithm Designer Module updates the cost metrics as information is provided by the Dependency Analysis Module, including information on cost metric completeness. A cost metric may not be considered complete if the dependent data structures have not been sufficiently selected so as to give an accurate cost estimate. For example, an algorithm timing cost metric may be dependent upon the wavefront data structure, which is unselected. As the wavefront data structure is selected and refined, the range of the algorithm cost metric is changed to reflect the actions of the Data Structure Designer Module. The cost metric is considered complete when the range of the algorithm timing cost metric is close enough to cause an algorithm to be selected (see rule ALGORITHM_SELECTION_refine_search_phase in the following section).

6.4.2. Algorithm Selection Operation

Figure 6-13 describes the three types of decision steps needed to complete the two tier algorithm selection process.

Try one of (in order):
1. Use domain knowledge to compose algorithm candidate sets and include dependencies to other algorithms.
2. Use domain knowledge to prune candidate set elements.
3. Select highest ranked element in the candidate set for some *arbitrary* router sub-task, if multiple elements remain; if only one remaining element, select that one.

Figure 6-13: Algorithm Selection Decisions

Refinement continues until either the router is completed, and there are schema selections for all necessary router sub-tasks, or no more information can be obtained either from the domain knowledge inference rules, or data structure refinements. In the latter case, the Algorithm Designer Module selects an arbitrary best candidate from an active candidate set, and then attempts to continue with refinement.

Only one of the three steps is performed by the Algorithm Designer Module during each iteration. The Algorithm Designer Module first attempts to *compose* candidate sets by adding candidates to unselected candidate sets using domain knowledge and previous algorithm selections. Not all decisions relating to task and subtask candidate set initialization are performed by the Input Stage. As algorithm selections are made, other tasks and subtasks may be created, or modified. The search and backtrace design tasks are especially susceptible to this type of change. For example:

```
ALGORITHM_SELECTION_search_restriction:
    Given minimizing execution time AND
    Given best-first expansion search algorithm THEN

    Propose a bounding-box limiting search as a
        search phase modification.
```

This example used router domain knowledge that most "well-routed" nets do not meander much outside a bounding box drawn around their net terminals in the routing grid. Hence, an appropriately sized bounding box can be used to exclude some parts of the routing grid from consideration, and limit the size of the expansion search space, and therefore the expected execution time of the expansion phase.

If no candidate scheme can be added to any candidate set, the Algorithm Designer Module attempts to prune candidates using domain knowledge and previous algorithm selections. For example:

```
ALGORITHM_SELECTION_refine_search_phase:
    Given minimizing execution time AND
    Given Dependency Analysis statistics
        for expansion decision AND
    Given time statistics are "close" THEN

    Select search phase option with best
        execution time statistics.
```

This rule combines the user directive to minimize execution time with expected execution statistics to select the best alternative. Note that the space consumption characteristic derived by the Dependency Analysis Module is ignored. When the user directs ELF to minimize execution time, ELF attempts to use the execution time metrics alone to make a decision. But if the execution time metrics are sufficiently *close* for several candidates, ELF uses the anticipated space consumption metrics to refine algorithm candidates. This situation is reversed when the user directs ELF to minimize space consumption, i.e., ELF only looks at the time estimates if the space consumption metrics are close, usually with 10% of each other.

Finally, if domain knowledge can not cause any additions to, or pruning from, any algorithm candidate set, the Algorithm Designer selects an arbitrary candidate set and selects its highest ranked candidate. For example,

```
ALGORITHM_SELECTION_arbitrary_selection
   Given no selections yet made this iteration AND
   Given an unselected candidate set AND
   Given close timing metrics for each candidate AND
   Given user directive to minimize time THEN

   Select candidate with smallest best timing metric.
```

This rule randomly selects a candidate set with close timing metrics and selects the candidate with the smallest best timing estimate from the Dependency Analysis Module.

After performing any of these steps, the Algorithm Designer Module incorporates the results of this selection into the ranking of the unselected candidate sets. For example:

```
ALGORITHM_SELECTION_reorder_candidates:
  Given timing of candidate A < candidate B AND
  Given no router domain knowledge can be applied to
        either candidate A or candidate B AND
  Given candidate B is currently preferred AND
  Given minimizing routing time THEN

Prefer candidate A over candidate B.
```

This rule reorders the ranking of the two candidates A and B within a single candidate task set. Each of the two candidates has been proposed as a solution to this particular task set. Before reordering, it is necessary to ensure that no *router domain knowledge* can be applied to either candidate which might change the timing costs derived by the Dependency Analysis Module. This rule also assumes that there was an original ordering that was contrary to the timing metric. Reordering only occurs when necessary, through some application of *router domain knowledge*. In this way, hazards such as infinite looping between the application of domain knowledge and later reordering is avoided. Figure 6-14 shows the effects of this hazard. The initial order was {A preferred over B}. Some *router domain knowledge* caused a change in preference to {B preferred over A}. Applying the result of cost metrics caused a change back to the original preference. If the same *router domain knowledge* is allowed to reorder the candidate preference set, an infinite loop occurs.

Figure 6-15 shows the types of domain knowledge used to implement the three steps making up the algorithm selection process used in the Algorithm Designer Module.

This section reviewed the algorithm selection process used in the Algorithm Designer Module, including the utilization of knowledge sources and the

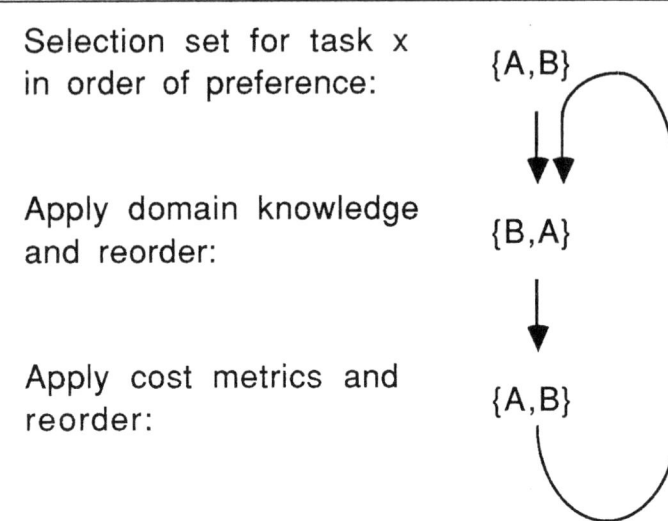

Figure 6-14: Selection Hazards: Premature selection

internal algorithm representation.

6.5. Chapter Summary

This chapter discussed the basic synthesis mechanisms of the Selection Stage and their implementations within the ELF architecture. The Selection Stage chooses the appropriate data structures and algorithms to implement the required routing task from candidates proposed by the Input Stage. The Selection Stage is composed of four modules: Selection Control, Dependency Analysis, Algorithm Designer and Data Structure Designer. The Selection Control Module controls the synthesis process while explicitly acknowledging the interdependent nature of data structure and algorithm synthesis. The Dependency Analysis Module analyzes the ADL candidate representations and formulates space and timing estimates for each representation based upon the

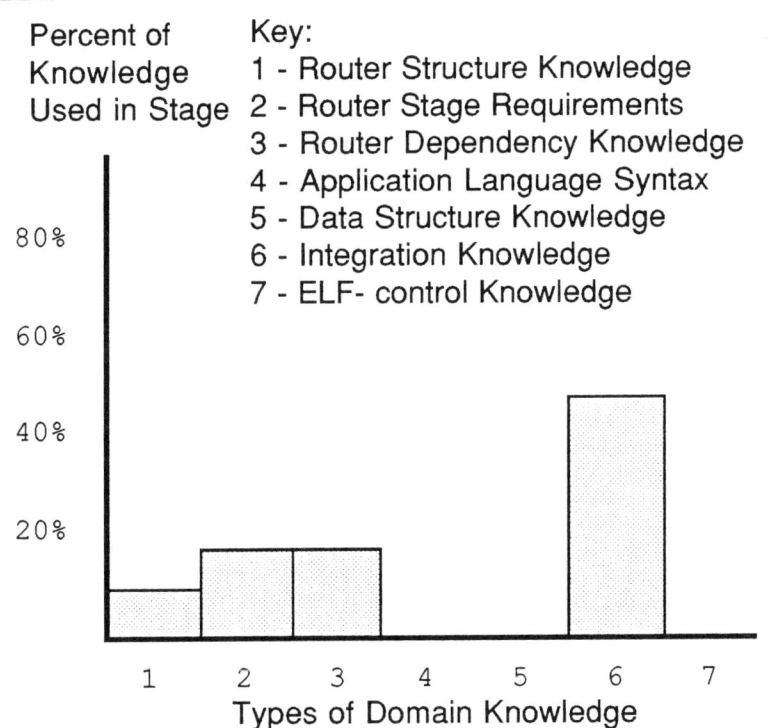

Figure 6-15: Domain Knowledge in the Algorithm Designer Module

current state of the synthesis process. The two Designer Modules iteratively select and refine their respective abstract representations until the final selections on all candidate sets are completed. The use of both *program synthesis*, *design generation* and *design interaction* knowledge drives the selection process. After the Selection Stage is finished, there are specific selections for all the algorithms and data structures required for each task the router must perform. These selections, still in abstract form, are passed onto the next stage, the Code Generator Stage, where they are transformed into executable C code. This process is described in the next chapter.

References

[1] J.D. Ullman, *Principles of Database Systems,* Computer Science Press, 1980.

[2] N. Wirth, "Program development by stepwise refinement", *Communications of the ACM*, ACM, April 1971.

Chapter 7
The Code Generator Stage

The output Code Generator Stage produces the executable code for the router. It takes as input the final candidate selections of algorithm schemas represented in ADL, and data structures represented as hierarchical compositions of simple templates. The Selection Stage produces these selections. Interestingly enough, the final candidate selections, while fully specifying a router at the design task and subtask level, do not completely specify an *implementation* of the router. The Code Generator Stage still must apply domain knowledge during transformation and synthesis of appropriate algorithms to produce the desired router code. The code generation process is based on a transformational approach, in which a sequence of stepwise refinements [5] ultimately lead to executable code, in this case, code written in C. The transformation process uses *design generation knowledge*, *design interaction knowledge*, and *program synthesis knowledge*.

There are the three key features of the ELF transformation engine:
1. The use of *router domain knowledge* to synthesize I/O operations.
2. The use of design generation and interaction knowledge as a guide within the transformation process,
3. The breakup of the transformation process into a sequence of stepwise refinements.

Figure 7-1 shows the type of domain knowledge utilized in the Code Generator Stage. In addition, this chapter describes an example illustrating a transformation sequence from ADL to the C programming language.

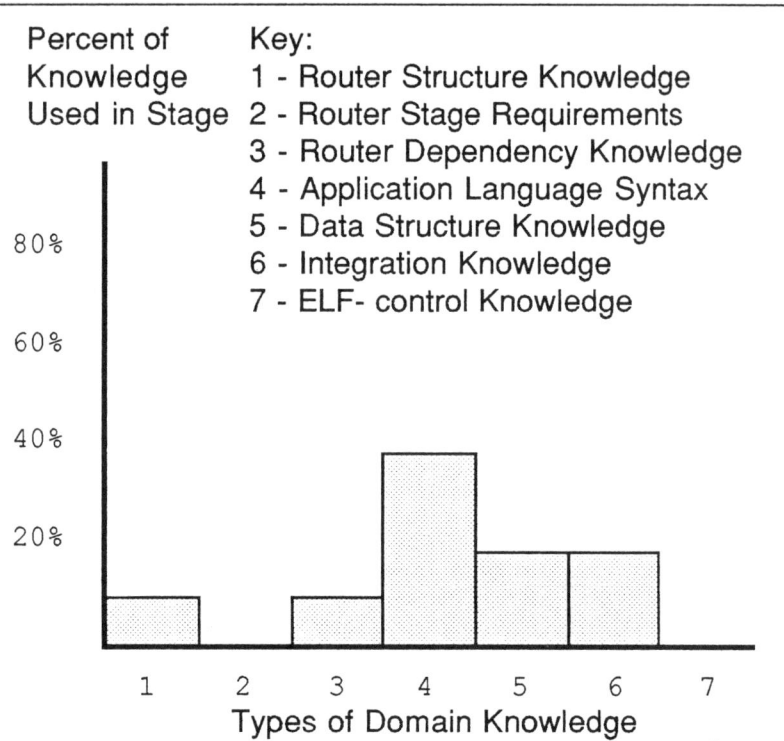

Figure 7-1: Types of Domain Knowledge in the Code Generator Stage

7.1. I/O Operation Synthesis

The first key feature is the use of *router domain knowledge* to synthesize I/O operations. These operations are either completely synthesized or modified within the Code Generator Stage using *router domain knowledge*. They are:
    1. Input of netlist

2. Output of routing paths

7.1.1. Input and Output Specification and Operation

The netlist input and routing path output subtasks are completely synthesized within the Code Generator Stage using standard compiler techniques. They are synthesized from a user-supplied BNF-like description. Parameters in this description define a model of the desired input (or output) netlist. The BNF-like format has two sections. The first section, the **define** section, tailors the netlist input task to the actual numerical parameters present in the input netlist file. For example: each record in the input file may be identified by a set of keys. The input file net records may be identified by specific key values. These values must be defined in the **define** section. The second section, the **model** section, describes the functionality of the input file in a BNF-like format. Although synthesis of these input and output tasks is fairly straightforward, it is nevertheless a critical part of ELF. For ELF to be a practical router generator, it must be able automatically to read and write the widely varying, unstandardized file formats used by real industrial routers. This simple BNF-to-code compilation provides the capability to handle a reasonably wide variety of formats.

7.1.2. Input Netlist Code Generation: An Example

In Figure 7-2, the first BNF statement in the **define** section organizes the input netlist file into a set of records. The second BNF statement organizes each record into an identification block, and connective information block. The next BNF statement specifies the order of the net record identification key values. Finally, the last BNF statement details the the format of the connectivity information. Altogether, four BNF statements were needed to specify the desired netlist format. The output netlist may be similarly defined.

```
DEFINE
     key1  0007
     key2  0001
     EOF   9999
MODEL
     FILE <= records EOF
     records <= id connective
     id <= key1 key2
     connective <= NET_NUMBER NODE_POSITION *
END
```

Figure 7-2: Sample BNF input file description

The BNF synthesis machine contains knowledge of the following keywords specific to the routing domain. They are: FILE, NET_NUMBER, NODE_POSITION, *, and +. The keyword FILE is the final BNF result and is therefore not used to synthesize the input routine. The keyword NET_NUMBER refers to the net numbering scheme used by the input netlist format. The following rules recognize the keyword NET_NUMBER and then produce the code required to read and check the two keys before reading NET_NUMBER.

```
NETNUMBER_recognize:
    Given bnf_parser token is NET_NUMBER AND
    Given right-hand-side token and call it RHS THEN

    Place NET_NUMBER as next child of RHS.
    Mark RHS as containing NET_NUMBER.

NETNUMBER_generate:
  Given terminal contains NET_NUMBER AND
  Given parent of terminal is defined by
     define section AND
  Given all definitions making up parent terminal THEN

  Build code to test for each key in order.
  Build code to read in NET_NUMBER.
  Build code to end each test for each key in order.


BNF_test_code:
   Given need to build test code against some value AND
   Given value has definition in define section THEN

   Generate:
       fscanf(fo, "%d", &value);
       if ((value == definition)) {

BNF_readin_code:
   Given need to build read in code for some value THEN

   Generate:
       fscanf(fo,"%d", &value);
```

The BNF-parser builds a parse tree of the parent-children for each BNF-like statement. The first rule, NETNUMBER_recognize, recognizes NET_NUMBER and places it as a child-terminal to its parent terminal, labeled *connective*. When the parser tree is completely built, ELF generates the code for each terminal in a top-down manner. Terminals consisting solely of definitions are bypassed until their meaning can be deduced from other terminals. In this example, the code to generate the key id's is tied to its parent's sibling's terminal, i.e., the generation of the code to read in NET_NUMBER, as in rule NETNUMBER_generate. Rule BNF_test_code generates a test for each key with its defined value given in the **define** section before the code to read the NET_NUMBER is generated in rule BNF_readin_code.

ELF assumes nets are numbered, but not necessarily numbered consecutively. The NODE_POSITION keyword identifies the appropriate physical placement of each terminal within the routing domain. The NODE_POSITION is a sibling terminal to NET_NUMBER and is generated next. The physical constraints of the routing domain identify the number and type of identifiers necessary to specify the NODE_POSITION keyword. All data structures dependent upon the NODE_POSITION, i.e., the internal structures representing all terminals in each net, are initialized to the physical mapping of these values to the internal representation. This is equivalent to going from a carrier definition of net with a terminal at point (x,y), to the actual cell indexed by (i,j) in the routing grid data structure. In the example shown in Figure 7-2, a gridded gate-array with 50 mil spacing between each internal router grid point, each input value must be divided by 50 to derive the internal grid point representation. Figure 7-2 produces the C code shown in Figure 7-3. The keywords * and + have their normal BNF-like meanings: 0 or more duplications and 1 or more duplications, respectively.

Automatic synthesis of input and output file formats allows ELF-produced software to work within already formed systems. In addition, automatic synthesis of input and output file formats allows ELF to be tailored to fit into a variety of new systems.

```
   gate1 = 0;
/* EOF check                          */
   while ((cellid1 != 9999 )) {
/* look for first id key              */
   fscanf(fo, "%d", &cellid1);
   if ((cellid1 == 7 )) {
/* look for second id key             */
   fscanf(fo, "%d", &cellid2);
   if ((cellid2 == 1 )) {
/* read in net identification         */
   fscanf(fo, "%d", &net_get1);
/* read in NODE_POSITION and put onto grid */
/* grid defined by user input template    */
   fscanf(fo, "%d %d", &position1,&position2);
   position1 = position1/50;
   position2 = position2/50;
   cells[gate1].x_cells1 = position1;
   cells[gate1].x_cells2 = position2;
/* build internal structure:
                     nets composed of cells */
   cellptr_get1 = (struct cell_on_net_def*)
      malloc(sizeof(struct cell_on_net_def));
   cellptr_get1->cell_on_net_next= NULL;
   cellptr_get1->cellnum_cell_on_net1 = gate1;
   gate1 = gate1+1;
/* read in 2nd NODE_POSITION, two point net
                              input only */
/* information supplied by input template */
   fscanf(fo, "%d %d", &position1,&position2);
   position1 = position1/50;
   position2 = position2/50;
   cells[gate1].x_cells1 = position1;
   cells[gate1].x_cells2 = position2;
   cellptr_get2 = (struct cell_on_net_def*)
      malloc(sizeof(struct cell_on_net_def));
   cellptr_get2->cell_on_net_next= NULL;
   cellptr_get2->cellnum_cell_on_net1 = gate1;
   gate1 = gate1 + 1;
   }}
   }
```

Figure 7-3: BNF to C Code Result (Comments Manually Added)

7.2. The Use of Router Domain Knowledge in the Transformation Process

The second key feature of the Code Generator is the use of *router domain knowledge* within the transformation process. In addition to *program synthesis* knowledge, the Code Generator Stage relies on *design generation* and *design interaction* knowledge, for example, knowledge about detailed implementation tradeoffs, and knowledge about how some candidate choices manifest themselves in low-level code. This knowledge, used to guide the transformation process, can be classified into three basic types along the same lines as the router input specification types provided in Section 5.1:

1. **Application Domain Knowledge:** concerns what *type* of router is required. Such knowledge is useful, for example, in determining *how* the expansion algorithm can best be tailored to meet such goals as limiting the expansion search within a bounding-box region.

2. **Algorithm Domain Knowledge:** links high-level choices about search schemas, backtrace schemas, etc, with low-level decisions about the code that implements them. In the above example, such knowledge is useful in determining *where* the limit checking is needed (which is dependent on the selected search schemas), as well as which data structures will be affected by the imposition of limit checking.

3. **Fabrication Domain Knowledge:** links detailed technology constraints to the final output code. This knowledge is concerned with how the final code must reflect the physical dimension of the problem. This knowledge is useful in determining *what* changes should be made to the final output code. For example, if 45° wiring is allowed, domain knowledge would be used to generate the detailed code to expand grid cells along both 90° and 45° expansion paths, allowing the final router to find 45° wire segments.

7.2.1. Effects of Applying Domain Knowledge

There are two categories of effects caused by the application of *router domain knowledge* to the transformation process. The first effect is the *redirection* of the transformation of an object, be it data structure access code or algorithm syntactic implementation. Redirection allows for the inclusion of code to handle algorithm modifications. The second effect is the *redefinition* of the transformable object. This redefinition may be at the task level, affecting all data structures and algorithms dependent upon this task; or it may be at the subtask level, affecting a single element within a data structure or algorithm (i.e., a change in the default initialization value for some element in a data structure to reflect the redirection of transformation). These two categories of effects are interdependent. Often, the redirection of the transformation process includes a redefinition of some transformable object.

7.2.2. Domain Knowledge Driven Transformation: An Example

A transformation redirection is made necessary by a Selection Stage decision to limit the expansion search with a bounding box. This decision manifests itself as a redefinition of the expansion search task to include the bounding box limitation subtask. A bounding box limits the allowable search space. This is shown in Figure 7-4. The figure on the left shows the search space without a bounding box. The figure on the right shows the much smaller search space with a bounding box in place. The use of a bounding box is based on the desirability of short routing connections. Search time is not wasted on areas unlikely to produce short connections.

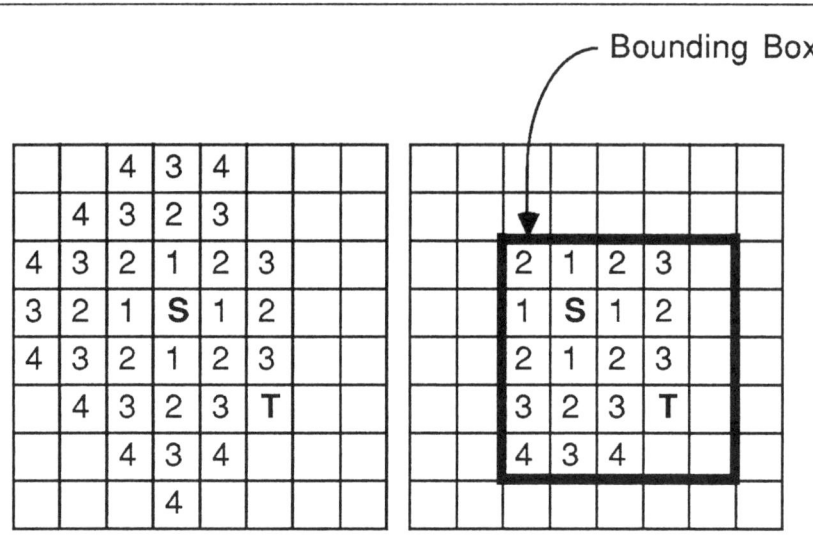

Figure 7-4: Effects of a Bounding Box on Search Area

The Code Generator Stage must modify the expansion search subtask to include the search limitation algorithm. The Algorithm Designer Module of the Selection Stage has already included the bounding box computation as a router subtask. The Code Generator stage must handle the synthesis of expansion search phase "hooks" to use the bounding box computation algorithm. Each "hook" must be placed within a known spot with the search algorithm. The Code Generator Stage has knowledge of the basic search algorithm framework. The Code Generator Stage synthesizes the expansion phase search algorithm using an internal algorithm *skeleton*. This skeleton defines each sub-task, e.g., define the start space, expand cell, and check if done, and allows the sub-tasks to be interconnected in different orders to produce different search types (e.g., depth-first, or best-first) [3]. This knowledge is exploited throughout the generation of the expansion search subtask. The skeleton is modified during the transformation process to reflect the code generation process.

In this case, the expansion search task description must now include additional constraints to redefine the limits of the search area. Domain knowledge must be supplied to coerce the transformation process to generate the appropriate code, for example:

```
GENERATOR_domain_application:
    Given transforming expansion AND
    Given a limits_LOOP condition AND
    Given the bounding-box method of
        limiting search THEN

    Modify the define search space framework slot to
      include a call to the bounding-box to
      derive the required limits.
    Modify the limits-check to check against
      these new limits.
    Modify the LOOP statement transformation
      to reflect these new limits.
```

The Code Generator Stage looks at the constraints on the expansion search phase skeleton and attempts to map them to algorithm and data structure modifications. For example, application knowledge tells the Code Generator that the *wavefront* data structure represents the search area. Thus, any modification to the definition of the search area has corresponding effects on the *wavefront* data structure. In addition, a change in the search area limits affects the *limits* data structure. This domain knowledge reference to the search area is combined with algorithmic knowledge of the search task operation to infer the necessary modification to the search algorithm. These three inferences are combined to derive a description of where the modification should occur in the search task to accommodate the additional subtask. This combination of inferences is an example of *Design interaction knowledge*. In this example, code generation modifies the expansion search algorithm framework to include soft limits checking within that part of the expansion skeleton responsible for defining the extent of the search space.

Algorithm knowledge about the functional requirements of the bounding box limitation subtask identifies what changes should be made. The Code Generator Stage is now directed to look for an iteration/test ADL sequence that references the *wavefront* and *limits* data structure within the search subtask. Identification of the correct sequence is assured by using domain knowledge about the functionality of the search algorithm. The following example uses algorithm domain knowledge to define the limits_LOOP statement used in the previous rule:

```
GENERATOR_define_limits_LOOP:
  Given transforming best-first AND
  Given a bounding-box to limit search AND
  Given best-first expansion search THEN

  Define limits_LOOP condition as ADL statement
    LOOP IN wavefront.
```

The expansion search algorithm must also determine which sites are expandable from a wavefront cell during the expand-cell subtask. A cell in the routing grid is *expanded* when it has been used to find its neighbors in the grid. The code to *find* these neighbors depends on the data structure choices (and is thereby sensitive to the target routing technology). For example:

```
GENERATOR_domain_fabrication:
    Given 45° wirings THEN

    Modify the allowable expansion set to
        include the 4 diagonal cell locations.
    Redefine the dimension of all variables
        dependent on the number of cell
        locations to reflect this increase.
    Define the default value for these
        new variables.
    Include these new variables.
```

In addition to determining where changes must be made to include diagonal wirings, code must also be generated to access the correct elements within the affected data structures. In this case, the grid data structure is accessed by the expansion subtask. The allowable expansion set locations must be mapped to access code. For example:

```
GENERATOR_expansion_sites:
    Given need to generate all cell expansion
        sites THEN

    Generate neighbors for routing grid.

GENERATOR_neighbors_array_array:
    Given generated neighbors for the grid AND
    Given an array of array grid
        implementation AND
    Given a row and column index for set THEN

    Define expansion sites to 4 sites:
            row index + 1,
            row index - 1,
            column index + 1, and
            column index - 1.
```

```
GENERATOR_neighbors_array_list:
  Given generator neighbors for the grid AND
  Given array of list grid implementation AND
  Given a row index for set AND
  Given column index is explicit   THEN

Define expansion sites to 4 sites:
      row index:    + 1,
      row index:    - 1,
      column index: set_next, and
      column index: set_last.
```

The first rule creates a subtask of the expansion subtask to generate all allowable neighbors. The second rule generates appropriate additional access code for the index variable for an array of array grid data structure (i.e., a standard two dimensional array treated as rows and columns of cells). The third rule generates appropriate additional code for an array of lists grid data structure (i.e., a non-standard one dimensional array of rows with linked-lists representing columns). These rules combine *router domain knowledge* of cell adjacency (fabrication knowledge) with the data structure implementation of the routing grid to generate the code to access the correct expansion sites.

The search algorithm transformation now includes the new expansion sites when transforming the expansion search space framework. This example shows both categories of effects caused by the application of *router domain knowledge* and *Design interaction knowledge* to the transformation process. The transformation process *redirects* the transformation process to include the bounding box algorithm and *redirects* the *limits* data structures to include the results of the bounding box algorithm.

In summary, *router domain knowledge* unravels the effects of decisions made in previous stages to assist in the current code synthesis task. Again, it is worth noting that ELF does *not* generate code by combining pre-designed, executable modules. Instead, the Code Generator of ELF is analogous to the code generator backend of a compiler: it transforms an abstract intermediate representation into executable code suitable for the target machine. *Router domain knowledge* drives the transformation process.

7.3. Stepwise Refinement in the Transformation Process

The third key feature of the Code Generator is the breakup of the transformation process into a sequence of stepwise refinements. *Program synthesis knowledge* generally drives the transformation process. The transformation process operates on ADL representation blocks. Each step in the refinement process transforms the ADL block from an ADL-text representation, to an intermediate format, before arriving at the final executable form. This section reviews the transformation process and then compares various transformations produced from a common ADL example.

7.3.1. The Transformation Process

As in other transformation systems, we must consider issues of the depth and breadth of a single transformation. Similar to the PSI-LIBRA system [4, 2], this stage generates final code as the result of a long sequence of many small transformations. Transformation starts from ADL and proceeds through an internal intermediate representation until the final code has been synthesized. Each step in the transformation process represents one change in either the statement syntax [1], or variable representation, or it represents the application of some knowledge source. The internal intermediate representation resembles a compiler parse tree (actually a compiler parse graph). In addition to elementary expertise about router code, the code generation process uses fundamental relationships between essential data structures and the desired output code to *amplify*, i.e., to translate and refine, the selected algorithm and data structures to the desired final executable form.

Transformation operates on syntactic blocks, from the innermost blocks outward. A syntactic block may be a simple statement, or a group of statements within loop statement markers. In the latter case, complete transformation of the loop may be dependent on how the interior loop statements are transformed.

A simple transformation sequence is shown in Figure 7-5. Initially, the statement shown at the top of the figure is purely in ADL. At Step 1, the code generator realizes the statement is a LOOP statement, with an embedded assignment statement. As expected, the Code Generator relies heavily on *program synthesis knowledge*. For example:

```
GENERATOR_assignment:
  Given ASSIGN token THEN

  Build ASSIGN statement in intermediate format.
  Type following variables.
  Check in LOOP block and if modifying within that
    LOOP, need to transform variable names.
```

In this rule, the Code Generator finds and builds an ASSIGN statement, breaks the statement into the intermediate form, checks if this statement has any effects on any statement higher up in the hierarchy, and prepares the ASSIGN

Automatic Programming Applied to VLSI CAD Software: A Case Study

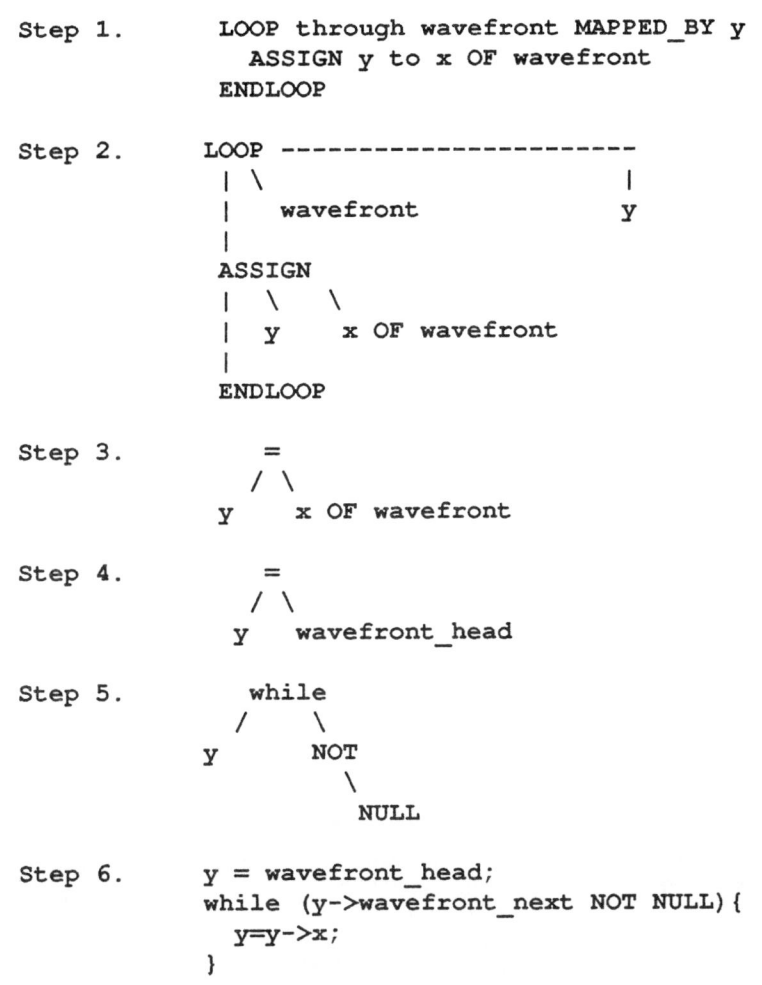

Figure 7-5: Code Generation Transformation Steps

statement for transformation.

The ADL block is transformed into the parse tree shown in Step 2. The LOOP statement is broken up into the object, wavefront, and the verb, y.

The data structure y is a verb because it causes an action on the set wavefront. Similarly, the ASSIGN statement is broken up into its right-hand-side and left-hand-side. Step 3 looks at the innermost block, the ASSIGN statement, and transforms the ASSIGN directive into a C equality symbol though the variables remain in their ADL format. In addition, step 3 notes that this assignment statement modifies the verb y that is defined higher up the block hierarchy. This information is supplied to the LOOP designer. The LOOP designer uses this information to suppress loop verb increment. For example:

```
GENERATOR_suppress_autoinc:
  Given in a loop block AND
  Given have auto-increment subtask for this loop
    block AND
  Given in left hand side, LHS, of ASSIGN
    statement AND
  Given LHS same as loop verb THEN

  Remove auto-increment subtask in this loop block.
```

In this rule, the code generator recognizes the left hand side of an assignment statement as being the reassignment of the loop verb. In this case, the loop block would not need an auto-increment at its end that would insure against an infinite loop.

The Data Structure Designer Module created two internal access variables from the variable wavefront, namely wavefront_head as the initial index to the chosen linked-list representation for the wavefront set and wavefront_next as the pointer to the next element in the set. The application of *router domain knowledge* caused the selection of the linked-list representation for the wavefront data structure. Step 4 utilizes these two bits of information to initialize the verb y to the initial index, wavefront_head for the `wavefront` set. For example:

```
GENERATOR_generate_head_of_list:
  Given loop set is a linked-list AND
  Given head of list variable for set, headoflist AND
  Given loop verb, verb THEN

  Generate before loop code:
        verb = headoflist;
```

Step 5 finishes the transformation of the actual LOOP statement using *program synthesis knowledge* of how to signify the ending of a linked-list representation.

```
GENERATOR_while_loop_list:
    Given loop block ready for synthesis AND
    Given loop set is a linked-list, listset AND
    Given listset has next element pointer, nextelement AND
    Given loop verb, verb THEN

    Generate:
        while( verb->nextelement NOT NULL) {
```

Step 6 shows the final C language representation of the ADL representation. *Router domain knowledge* was not applied in this example, other than in the case of the Selection Stage selection of a linked-list representation of the `wavefront` set.

7.3.2. Transformation Comparison

The transformation process is heavily dependent upon Selection Stage decisions and the user input specifications. This section reviews the transformation process of a common ADL sample, but for two different user input specifications. The sample ADL is shown in Figure 7-6. This sample loop iterates through each element in the set *grid* assigning the subelement *how_got_here* to the unique identifying variable, *iterator*. The following two experiments display different transformation processes and, correspondingly different synthesized code.

```
LOOP iterator IN grid END
  ASSIGN how_got_here OF grid MAPPED_BY iterator TO
    iterator END
ENDLOOP
```

Figure 7-6: Transformation Example: ADL

Figure 7-7 shows the relevant portion of the user input specification for example 1. ELF is directed to produce a global router for a gate-array with an emphasis on minimizing the execution time. These specifications, combined

```
Algorithm
        expansion
                minimize        time
        done
Application
        type                    global
        sub_type                gate_array
```

Figure 7-7: Example 1: User Input Specification

with the fabrication specification indicating a representable gate-array size, cause the Selection Stage to select a two-dimensional array for the *grid* data structure. ELF selects an integer representation for both the *iterator* and *how_got_here* variables.

The transformed C code using these input specifications as a guide is shown in Figure 7-8. The LOOP statement, with its embedded ASSIGN statement, is then relatively straightforward. First, the ASSIGN statement is redirected to use the internally-generated variables *iterator1* and *iterator2* in place of *iterator* so as to specify a unique element in the *grid* data structure. The ASSIGN is also redefined to include both dimensions of *how_got_here*, represented as *how_got_here1* and *how_got_here2*. This results in the synthesis of two C assignment statements. Similarly, the LOOP is redirected to use the same internally-generated variables, a for C loop construct is used to iterate through the two-dimensional array chosen for the *grid* data structure. Nested for loops are a result of LOOP redefinition to include both dimensions of *grid*.

```
for (iterator1=0;iterator1<25;iterator1++) {
   for (iterator2=0;iterator2<36;iterator2++) {
      grid[iterator1][iterator2].how_got_here1
         = iterator1;
      grid[iterator1][iterator2].how_got_here2
         = iterator2;
   }
}
```

Figure 7-8: Example 1: Sample Code

Figure 7-9 shows the relevant portion of the user input specification for example 2. ELF is directed to produce a graph-based global router for a macrocell IC. These specifications cause the Selection Stage to select a graph representation for the *grid* data structure. Each node has an adjacent-node list. A graph is a multi-rooted list (tree) of lists. ELF selects a graph-node pointer representation for the *iterator* variable, but selects an integer representation for *how_got_here* based on *router domain knowledge* identifying *how_got_here* as uniquely identifying the physical location of an element. In addition, an

```
Application
        type              global
        alg_type          graph
Fabrication
        technology        IC
```

Figure 7-9: Example 2: User Input Specification

adjacency variable defining the adjacent physical regions is declared and a list of graph-node pointers is chosen as it representation.

The transformed C code using these input specifications and Selection Stage decisions is shown in Figure 7-10. This example is considerably more complex than in Example 1. First, on the left-hand-side, the embedded ASSIGN statement has *iterator* indexing into *grid*. Since *iterator* is a pointer to a *grid* terminal, the subelement *how_got_here* can be identified directly from *iterator* itself. On the right-hand-side, the ASSIGN statement also has the *iterator* variable. *iterator* uniquely identifies a terminal within the *graph* data structure, but it does not identify the physical location of each terminal. In the first experiment, *iterator* fulfilled both functions. In this case, the function of identifying the physical location is subsumed by the internally-generated variable *position_grid*. This redefines the transformation process to instead access the *position_grid* variable as a subelement of the *iterator* variable.

The LOOP transformation is redirected to use a recursive search as the iteration vehicle. The recursive subroutine is defined to include the ASSIGN statement with additional code to reach, via recursive calls, all of its adjacent

Automatic Programming Applied to VLSI CAD Software: A Case Study

graph nodes. Domain knowledge of the *adjacency* data structure and *program synthesis knowledge* of how to iterate through the chosen list implementation result in a while loop with automatic increment via the internally declared *adjacency_next* pointer. In this way, the recursive search "iterates" through the entire set *grid* and assigns both dimensions of the variable *how_got_here* to the identifying *position* value.

```
        iterator = grid_head;
        inloopgrid1(iterator,marker_value);
        marker_value = marker_value + 1;

        ...

        void inloopgrid1 (loopvar,marker)
        struct grid_def *loopvar;
        int marker;
        {

          loopvar->how_got_here1
                  = loopvar->position_grid1;
          loopvar->how_got_here2
                  = loopvar->position_grid2;
          loopvar->internal_mark = marker;

          nextone = loopvar->adjacency_head;
          while (nextone <> NULL ) {
            if (nextone->internal_mark <> marker) {
              inloopgrid1(nextone,marker);
            }
            nextone = nextone->adjacency_next;
          }
        }
```

Figure 7-10: Example 2: Sample Code

The two examples demonstrate the effect of decision-making, using either *router domain knowledge* or *program synthesis knowledge*, on the transformation process.

7.4. Chapter Summary

This chapter discussed the basic issues of the Code Generator Stage and its implementation within the ELF architecture. This stage uses the program synthesis technique of transformation to produce executable C code from ADL algorithm representations. *Design generation knowledge*, *design interaction*, and *program synthesis knowledge* drive the transformation process. With this background, we can now proceed to survey ELF's actual implementation.

References

[1] A. Aho, and J. Ullman, *Principles of Compiler Design,* Addison Wesley, 1978.

[2] D. Barstow, "An Experiment in Knowledge-based Automatic Programming", *Artificial Intelligence,* Vol. 12 1979, pp. 73-119.

[3] E. Horowitz and S. Sahni, *Fundamentals of Computer Algorithms,* Computer Science Press, Inc., 1978.

[4] E. Kant, "On the Efficient Synthesis of Efficient Programs", *Artificial Intelligence 20* 1983.

[5] N. Wirth, "Program development by stepwise refinement", *Communications of the ACM*, ACM, April 1971.

Chapter 8
Implementation

This chapter discusses some implementation characteristics of the ELF system, a brief design history and the procedure to modify ELF when some facet of the target technology expands beyond ELF's current range of supported technologies.

8.1. Implementation Characteristics

ELF itself is a rule-based program written in OPS5 [1], running under UNIX^{tm} 4.3. The routers produced by ELF are written in C. A rule-based implementation is appropriate in programming situations where the domain information can be simply expressed as a set of situation-action pairs, and where there is an expectation that the codified information will grow as the system is developed. Our experience to date suggests that *router domain knowledge* is conveniently represented as rules. Moreover, the transformation approach used in the Code Generator is most naturally implemented as a large body of specific rules. For simplicity, almost all of ELF (except some basic I/O) is written in OPS5, even though there are some algorithmic subtasks in ELF that could more conveniently be written in a conventional language.

ELF currently contains 1323 rules, which can be broken down by stage or by

Stage	Number of Rules	Router Domain Rules	Program Synthesis Rules	Integration Knowledge Rules	Control/ Misc. Rules
Input	76	17	0	0	59
Selection	574	136	169	186	83
Code Generation	673	4*	419	112	136
Total	1323	157	588	298	278

* Does not include any domain knowledge driven rule building

Figure 8-1: ELF Rule Breakdown By Stage

content--*router domain, program synthesis, design interaction knowledge,* and control--as shown in Figure 8-1. Each stage uses both domain and program synthesis knowledge, though in different amounts. The use of *program synthesis knowledge* increases and the use of *router domain knowledge* decreases as synthesis proceeds. Control rules handle the ordering of events, from reading in the input, to selection of the necessary data structures and algorithms, to formatting the output code. Figure 8-1 does not include any domain knowledge rule building. These rules are explained in the following section.

8.1.1. Input Stage

The Input Stage has *router domain knowledge* rules. These rules guide the user-specification inference mechanism and produce the relevant set of router tasks and subtasks. Examples are found in Chapter 5.

8.1.2. Selection Stage

How knowledge in the implementation of the Selection Stage is distributed across its respective modules is shown in in Figure 8-2. In most Modules and Stages, *program synthesis knowledge* refers to knowledge of C program syntax and constraints, as well as implementation specifics. An exception is the Dependency Analysis Module, which has knowledge of the ADL program syntax and constraints. Both language-specific knowledge sources are classified as *program synthesis knowledge*.

As expected, both the Algorithm Designer and the Data Structure Designer integrate both sources of knowledge, but each knowledge source is most heavily concentrated in the appropriate Designer: *program synthesis knowledge* is concentrated in the Data Structure Designer Module and *router domain knowledge* is concentrated in the Algorithm Designer Module. In addition to pre-defined *router domain knowledge* rules, the Data Structure Designer Module also builds appropriate *router domain knowledge* rules to fit the desired application. For example, the Data Structure Designer Module *builds* rules that allow the Code Generator Stage to automatically recognize and then act on the data structure manipulation techniques required in a search design modification. Specifically, the Data Structure Designer Module may build rules to handle specific special case transformations, e.g., the transformation of a loop into a recursive subroutine call. The Algorithm Designer Module also contains more control-type rules to handle the increased complexity of algorithm cost computation, with all its embedded data structure cost analysis.

Selection Stage Module	Number of Rules	Router Domain Rules	Program Synthesis Rules	Integration Knowledge Rules	Control/ Misc. Rules
Selection Control	12	0	0	0	12
Depend. Analysis	72	0	66 +	0	6
Data Structure Designer	182	16*	26	125	15
Algorithm Designer	308	120	77	61	50
Total	574	136	169	186	83

* Does not include router domain knowledge-driven rule building:

+ ADL parsing program synthesis knowledge

Figure 8-2: Selection Stage Rule Breakdown by Module

8.1.3. Code Generator Stage

Knowledge placement in the Code Generator Stage is shown in Figure 8-3. As expected, the simple transformation task accounts for the bulk of the program synthesis rules. *Router domain knowledge* and *program synthesis knowledge* are required for the more complex redirection and redefinition of the transformation process. The control and miscellaneous rules handle the ordering of transformation events and output of the final code.

Code Generator Stage Tasks	Number of Rules	Router Domain Rules	Program Synthesis Rules	Integration Knowledge Rules	Control/ Misc. Rules
Variable Declaration	73	0	57	13	3
Transformation	413	4*	285	72	52
I/O Generation	127	0	77+	27	23
Output	58	0	0	0	58
Total	673	4*	419	112	136

* Does not include router domain knowledge-driven rule building:
+ ADL-parsing program synthesis knowledge

Figure 8-3: Code Generator Stage Rule Breakdown by Task

8.2. Design History

Figure 8-1 shows that about half of ELF's total rules are in the Code Generator. This is to be expected in a system of this type, since ELF was built bottom up. That is, the Code Generator is the oldest and most mature part of ELF. Early in the design of ELF, it was decided to formalize and implement the candidate representations and transformations to executable code, and use this as the foundation for experimenting with the more abstract problem of selection. Hence, a functional Code Generator, with rather simple Input and Selection modules formed the first operational version of ELF [2]. This version relied more on generic *program synthesis knowledge* than on *router domain knowledge*. As ELF evolved, its Input Stage and Selection Stage then grew, primarily by the addition of *router domain knowledge*. In addition, the Code Generator Module grew slightly as new ADL algorithm schemas were added, and as new low-level code-writing problems were attacked. Recent experience has shown that as ELF stabilized, almost all of the growth in the rule base has been in the Selection Stage.

8.3. Modifying ELF: Is It Really Better?

One of the prime motivations behind ELF was to improve upon the current solutions for managing large technology-sensitive software systems. The most common solution is to devise "general purpose" tools that, in some form, include code to address the myriad special cases that may arise when the tool is configured for, and then applied to a wide range of target applications. The critical problem is in devising sufficiently general software architectures so that *all* new cases can be accommodated without major software redesign.

Moreover, such general tools usually carry all this baggage in their runtime environments--whether they need it or not for their current application--which potentially compromises the efficiency of the tool as compared to its custom designed counterpart. The ELF architecture addresses this problem in a fundamentally different way. We still consider all the "cases," but in a knowledge-based *generator* for the tool, not the tool itself. One of our goals has been to organize domain knowledge to facilitate the management of software redesign in rapidly changing environments. Hence, a critical question is: how hard is it to modify ELF itself? To answer this, in this section we illustrate what changes to ELF are necessary when a new target application evolves.

8.3.1. How To Add a New Technology?

In this section we show how to modify ELF to handle a new technology. (Currently, all modifications are done by hand, however, we anticipate that rule regularity could be exploited by a knowledge acquisition tool.) Specifically, we describe the number and type of additional rules that would be required to handle a new technology. The example we will use is the addition of diagonal (45°) wires to the traditional rectilinear wires.

There is generally no need to write new algorithm schemas when adding a new technology. In addition, the current support for data structure types and composition is robust enough to handle the stress of adding diagonal wires, so no new data structure types are needed.

However, adding diagonal wiring does represent a change in the allowable fabrication constraints read by the Input Module. The Input Module will need to modify the input parsing rules to read in the new specification. In addition, it is necessary to write one new rule to handle wiring incompatibility constraints that arise from this new fabrication possibility. Specifically, this rule must note that additional expansion sites can be reached in a single step in the expansion search sub-task.

There is no need to change the Selection Control Module, Algorithm Designer, or Dependency Analysis Module. However, the Data Structure Designer will need one new rule for each of the existing rules used in the expansion search sub-task that define the expansion site variable to access expansion sites in the grid. The new rules deal with the new cell adjacencies as expansion sites implied by 45° wires. For example, for a global router, the grid data structure must provide a measure of diagonal wiring capacity between adjacent cells. The Data Structure Designer is robust enough to manage the impact of diagonal expansion on all data structures used in the expansion search sub-task. Given these new rules, the space and time requirements for these data structures will be computed correctly.

The Code Generator will need one rule for each root data structure type--array or list--to provide the correct mapping from domain ideas about diagonally adjacent routing cells down to the specifics of generating the correct code to access the new expansion sites. Finally, one rule is necessary to map domain ideas about a bounding box to limit the cell search space down to the mechanics of the limit-checking code. Given these rules, the Code Generator Module will automatically generate the correct expansion search necessary to handle diagonal wires given these actions.

In the next chapter, results from a router generated to attack this new technology will be presented. Adding diagonal wiring to the fabrication constraints caused the addition of 5 rules. Altogether, less than 100 new lines of OPS5 were added to the ELF system to handle the new fabrication process.

8.3.2. How To Add a New Algorithm Representation?

Another way technology can evolve is the development of new routing algorithms or new routing representations. In this section, we show how to modify ELF to handle a completely different routing representation. Specifically, we show how to include a graph-based routing algorithm.

So far, we have only discussed maze routers that route wires on a rectangular grid of identical cells. However, other representations are used as well. For example, when the modules being wired are of large and irregular shape (as with the component parts of a custom IC), the regions between the modules for wiring are themselves irregular. If we again represent each routing region as a cell in the routing process, a graph structure is most amenable. Regions are graph vertices, and edges represent region adjacencies. Analogous processes for cell expansion, backtrace and cleanup can be performed on this data structure [3].

There is no need to write new router algorithm schemas for the basic component phases of setup, expansion and cleanup, because these maze routing algorithms already included in ELF are sufficiently general to handle graph-based routing. However, the backtrack phase is sufficiently different so as to require a new ADL schema representation. The current support for data structure types and composition is strong enough to handle the required graph structures.

This new algorithm forces a change in the allowable algorithm constraints to the Input Module. The Input Module will need to modify one existing rule to read in the new GRAPH algorithm constraint. In addition, it is necessary to write one new rule to handle new constraints that arise from interactions with this new algorithm constraint (e.g., an interaction between the graph algorithm requirement and the granularity of the routing space data structure with respect to the physical constraints) and one new rule to infer this constraint if the user specifies only that they want to route a custom IC.

There is no need to change the Selection Control Module or the Dependency Analysis Module. However, the Algorithm Designer will need a rule to recognize the indefinite number of expansion sites available from a cell, rather than the fixed number in previous examples and the corresponding affects of search modification. In addition, a rule must be added to force the backtrace algorithm to reference the graph structure directly.

The Data Structure Designer will need a rule to handle the new effects on both the wavefront structure and the grid structure. Specifically, *router domain knowledge* should indicate the indefinite number of adjacencies for each element in these structures. A new data structure to handle the organization of the data structure adjacencies must be defined. The Data Structure Designer is robust enough to favor list implementations in unbounded, e.g., indefinite, situations such as these. In this particular case, each cell will contain a list of all adjacent cells. The expansion cost function representation also becomes more complicated. The expansion cost function must reflect the physical nature of the problem, i.e., the cost to reach each adjacent cell has two components: the distance-based cost to traverse the relevant portion of the cell we are expanding and the congestion-based cost to enter the adjacent cell we are trying to reach [3]. Each expanding/adjacent cell pair has different traversal and entry costs and each cell must store this information. A rule to generate the necessary variables, and to generate the correct cost function will need to be written.

The Code Generator will need a rule for each expansion search algorithm to recognize the adjacent cell list and generate correct code to access each element in the list. Depending on the expansion search style, each new rule might modify the expansion search sub-task and allow multiple visits to a cell as expansion proceeds; this simply reflects the nature of the search process on this data structure. The Code Generator will automatically act on the search modifications set up in the Selection Stage.

In the next chapter, results from a simple router generated to attack this new graph-based algorithmic strategy will be presented. In summary, about 20 new rules were added, and one new algorithm schema was written. Altogether, we estimate approximately 300 new lines of OPS5 and 89 new lines of ADL (to handle the backtrack phase) were added to the ELF system to handle simple graph-based routing.

8.4. Issues in Debugging ELF-synthesized Code

Unfortunately, the ELF synthesis architecture does not guarantee correct construction of the synthesized code. More precisely, ELF does not provide a proof of the transformation mechanism and so ELF-synthesized code may fail to compile, execute, or produce a reasonable routing solution. Such failures may be classified under two main types of failures. They are:

- **Program Synthesis Failure**: ELF fails to correctly produce executable code. Examples of this type of failure range from variable access errors such as failing to dereference a pointer when required, to infinite loops. This failure is due to incorrect program synthesis knowledge.
- **Domain Knowledge Failure**: The ELF router fails to produce a reasonable routing solution. Examples of this type of failure range from incorrect search modification due to immature domain knowledge, to incorrect task decision-making. In our experience, this failure is usually due to immature domain knowledge.

As an example, we consider the graph-based algorithm addition described earlier. We detail the steps that were required to debug ELF itself when we actually added these new rules, and to debug the ELF-synthesized routers, until ELF produced correct graph-based routers.

We uncovered three major errors in this debugging process. The first failure was in the modification of the search phase task to accommodate the new adjacency definition. This was a direct result of a lack of domain knowledge. ELF failed to capture the subtleties of the depth-first expansion scheme within a graph-based algorithmic approach. The second failure was also in the modification of the search phase task. ELF failed to correctly access the new cost function. This was a result of a lack of integration knowledge. *Design interaction knowledge* should have marked the necessity of redirecting transformation when transforming the cost function to apply the new graph-based requirements. The correct domain knowledge was present, yet it was not applied. Finally, ELF failed to correctly pass pointer parameters. These pointer parameters should not have been dereferenced. This was a result of a lack of *program synthesis knowledge*. We fixed this problem by noting the interaction of design requirements for pointer parameters and the syntactic requirements of passing parameters. Two rules were added to redirect the transformation process using *design interaction knowledge* as a guide.

In general, ELF produces executable code for known applications. But, during the integration of new design knowledge, ELF may fail to capture all of the subtlies involved, and some debugging of ELF based on examination of its generated routers is usually required.

8.5. Chapter Summary

This chapter discussed some ELF implementation characteristics and the breakdown by rules of the various knowledge sources in each stage and module. ELF is a rule-based program written in OPS5, produces routers in C, and currently contains 1323 rules. Each stage of the ELF architecture uses both *router domain* and *program synthesis knowledge*. As synthesis proceeds, the use of *program synthesis knowledge* increases and the use of *router domain knowledge* decreases, though both are required throughout the synthesis process. In addition, this chapter described two preliminary examples explaining the effort required to modify ELF when technology or algorithms expand beyond ELF's current domain operating range. These examples demonstrate how ELF can evolve to meet some new target applications.

References

[1] L. Brownston, R. Farrell, E. Kant, and N. Martin, *Programming Expert Systems in OPS5,* Addison-Wesley, 1985.

[2] D. Setliff and R. Rutenbar, "Knowledge-Based Synthesis of Custom VLSI Physical Design Tools: First Steps", *Conference on Artificial Intelligence Applications*, IEEE, March 1988.

[3] J. Soukup and J.C. Royle, "On Hierarchical Routing", *Journal of Digital Systems*, Vol. V, No. 31981, pp. 265-289.

Chapter 9
ELF Validation

In this chapter, we demonstrate the feasibility of the ELF methodology, i.e., the feasibility of a synthesis architecture to transform high-level specifications to executable code. Of course, we note that the routers that ELF can currently produce are not production-quality: the ELF synthesis architecture is itself a prototype and it generates prototype-quality routers. We do not regard this as any limitation of the ELF synthesis architecture; the quality of ELF's routers reflects both the quality and quantity of router domain knowledge embodied in ELF. For this research, our goals have been to embed enough knowledge for ELF to build a wide range of basic, fully-functional routers, each sophisticated enough that it can be tested against a set of realistic synthetic and industrial routing benchmarks. But as we shall see, ELF has managed to build some routers that perform surprisingly well, even when measured against hand-crafted software and pathologically difficult benchmarks.

This chapter begins with a description of our overall experimental methodology for evaluating the ELF architecture, and then presents results from a series of different ELF-synthesized routers running a range of synthetic and industrial benchmarks.

9.1. Experimental Methodology

In Chapter 4, we recognized three classes of constraints driving the ELF synthesis process: application constraints, algorithmic constraints, and fabrication constraints. Validation of the ELF synthesis architecture requires the generation of a reasonable variety of routers each addressing a different set of these constraints. After building these routers, we must select benchmark tasks that realistically exercise each router. We select some adequate synthetic experiments and two industrial examples as benchmarks for our synthesized routers.

It is important to recognize that there are basically two different kinds of measures of success for ELF-generated routers. First, we must insure that simply as a program, the ELF-generated code is syntactically well-constructed and free of obvious errors. In short, we must determine if it will run *at all*. Second, and certainly more critical from the CAD viewpoint, we must determine how *well* it functions as a router, measured against the specifications it was synthesized to meet, and users' expectations about the quality of its final output wiring. The benchmarks answer the following questions, detailing the two measure of success for each ELF-generated router:

1. **Does it compile?** This tests the basic application language syntactic program synthesis knowledge base.

2. **Does it execute?** This tests the data structure implementation program synthesis knowledge base.

3. **Does it produce a viable routing solution?** This tests router structure, router dependency knowledge, and integration knowledge bases.

4. **Does it produce a viable routing solution in a timely fashion?** This tests the basic cohesive nature of all knowledge bases.

Successful compilation and execution are implicit in the successful reply to the third and fourth questions. To answer the question of producing viable routing solutions, we will present measurements of the routing quality in terms of typical wiring metrics like completions, i.e., percent attempted wires successfully routed, wirelength, number of vias, etc. To answer the question of execution quality, we present a number of statistics detailing the following:

- **Execution Time**: we measure the execution time to complete the routing task on a VAX 8800.
- **Memory Utilization**: we measure the amount of space in bytes required to complete the routing task. Because we know *a priori* the major contributors to memory allocation are the *grid* and *wavefront* data structures, we present data totaling the maximum dynamic memory allocation for these data structures during execution.
- **Comparison with hand-developed code**: we compare the above statistics with hand-developed code, where possible, to provide some concrete comparisons.

Finally, we present statistics detailing the performance of ELF itself in the generation of each router. We give the number of rules fired in each stage and the final Selection Stage decision set. In this way, we can summarize the gross differences between each generated router.

In the following sections, we describe six router-synthesis experiments which we partition into three different styles. The first style builds a suite of four global routers for gate array style semi-custom ICs. We use a single synthetic gate array benchmark to evaluate this suite of routers. The second style builds a router to be used for global routing of printed-circuit boards. Two industrial PCB benchmarks are used to evaluate this router. The third style builds a router to be used in a macro-cell IC application utilizing a graph-based (not grid-based) algorithm. A simple synthetic macro-cell IC benchmark is used to evaluate this router.

Together, the six experiments demonstrate the basic capabilities of the ELF system. This chapter summarizes results from these six router synthesis experiments, briefly analyzes what decisions were made by each stage in the ELF system and why, and provides performance statistics for the synthesized routers.

9.2. Gate Array Style Routers

These experiments stress the wide range of application, algorithm, and fabrication options supported by the current ELF synthesis architecture. The gate array style experiments consist of four routers to be used for global routing of grid-based gate array chips. Because this is a global routing task, individual cells in the routing grid represent large areas, basically pieces of *routing channels*, on the chip. The cost of traversing a cell increases as the wiring capacity decreases because previous wires that traversed the cell consumed scarce routing resources. These experimental routers must capture the salient features of this particular set of router design problems. Specifically, this set of gate array routers stress the domain knowledge-driven selection of the router expansion task and the supporting data structures. We summarize the critical features of each gate-array router experiment as follows:

- **Router 1:** Ordinary fabrication constraints, e.g., a two-dimensional grid of regions, some occupied by placed cells, some available for routing with a user-selected cost-function to match wiring demand to wiring capacity. Multi-point nets are routed directly, i.e., not decomposed into a sequence of two-point nets. ELF was directed to synthesize this router to optimize routing speed.

- **Router 2:** Same fabrication constraints as Router 1, but ELF was directed to synthesize to optimize (minimize) the required memory space.

- **Router 3:** Changed fabrication constraints to require that multi-point nets be decomposed in an optimal fashion (using a minimum spanning-tree algorithm [1]) into a set of independent two-point nets. In addition, ELF was directed to synthesize to optimize routing speed.

- **Router 4:** Hybrid fabrication scheme using coarser cells and allowing diagonal 45° wiring between these coarser cells. This constraint is actually more representative of a class of medium-grain large-scale PCB routers than it is of typical gate array routers, but we include it here for two reasons. First, it demonstrates the flexibility available in ELF for generally accommodating modifications of the underlying router technology *without* having to modify executable router code. And second, by applying it to this gate array problem, we generate a router that can be compared to the other three more traditional gate array routers. This is the 45° wiring capability whose addition to ELF was described earlier in Chapter 8. In addition, ELF was directed to minimize the required memory space.

These experiments place stress on the components of ELF responsible for the design of the expansion phase algorithms, and the design of the grid data structure that supports the expansion search process.

9.2.1. Comparison of ELF-synthesized Gate Array Routers

Each of Routers 1-4 was synthesized in about one hour on a VAX 8800. The generated routers varied in size from 1510 to 2051 lines of C code. Figure 9-1 gives the resultant size and number of ELF rule firings to synthesize each router. ELF also internally builds rules to handle special case transformation steps, e.g., the recognition of an iteration through a graph data structure and the subsequent transformation redirection to a recursive subroutine call. The number of internally built rules for each experiment is also shown. ELF averages about 16 rules to transform and produce a single line of C.

	Experiments			
	1	2	3	4
Lines of C Code	1510	1913	1632	2051
ELF Rule Firings				
Input	71	71	71	71
Domain Interpreter	7367	8445	7516	6685
Code Generator	23252	31375	26492	33490
Internally Generated	0	49	0	22
Total Rules Fired	30690	39940	34079	40052

Figure 9-1: Code Size and ELF Rule Firings for Gate Array Experiments

Specifications	Experiments			
	1	2	3	4
Fabrication	gate array	gate array	gate array spanning-tree	gate array diagonal
Minimize	speed	space	speed	space
Decisions				
Search	best-first	depth-first	depth-first	depth-first
Grid	array of array	array of list	array of array	array of list
Wavefront	array of list	array of list	array of list	list

Figure 9-2: Gate Array Experiment Decisions

Figure 9-2 summarizes the ELF decision process for each of the four experiments. With respect to choices for the style of the expansion phases, the first router chose a general-purpose best-first style. Since this router was not forced to address any unusual fabrication constraints, this is the obvious choice. The second router, however, chose a depth-first expansion style to minimize the space required by the router. A depth-first router produces a smaller wavefront, and as such meets the space minimization specification. The third router also chose a depth-first expansion style. Router 3 uses a depth-first style because ELF knows that it needs only to route two-point nets (recall that multi-point nets will be decomposed in Router 3). Depth-first schemes can search very fast and are most effective in practice when restricted to two-point net routing problems. Hence, ELF decides to take advantage of this wiring constraint and use this scheme. On the other hand, Router 4 also uses a depth-first scheme, but for a different reason. Since 45° wiring is permitted here, all cells in the routing grid have eight neighbor cells to be examined during expansion search. This is unlike the other routers, which need only examine the four compass neighbors, and can ignore the four diagonal neighbors. Because ELF understands that the number of cells to be examined strongly determines the running time of the router, and because this router appears as though it will examine *many* more cells, ELF is again inclined toward a depth-first style with the explicit aim of reducing the total number of cells visited during the expansion process. In addition, because the granularity of the cell grid was coarser for Router 4 (i.e., the mapping of internal grid structure to physical region is broader in area), ELF was again inclined to use a depth-first scheme since a coarser cell grid is likely to have fewer blockages, which can seriously degrade the efficiency of a depth-first scheme. Hence, ELF chose a depth-first scheme here as well.

With respect to memory usage, Routers 1 and 3 were optimized for speed, not space. Hence, in this case, ELF chose a two-dimensional array (an array of arrays) of records as the fastest implementation for the grid data structure. Because ELF knew this was to be used on routing problems of modest size, it ignored knowledge about the expected sparsity of this data structure (knowledge that would have suggested a list-based data structure) and chose the much larger array style to maximize speed. In contrast, Routers 2 and 4 were optimized explicitly for space. In this case, ELF chose an array of lists of records for the grid. This is a highly unusual data structure selection. The most typical commercial selection is a two-dimensional array. But because ELF was given no domain knowledge that it should **not** try something else, and because it knew this was a small routing problem, it built this novel data structure which, perhaps surprisingly, works fairly well. Since occupied cells are blockages through which no wires can be routed, and such cells occupy a large fraction of the gate array's areas, ELF simply decided not to represent them in the grid data structure. The resulting array of lists is essentially a sparse array structure, where ELF substitutes computation for storage whenever it needs to deal with these cells. During selection of the root style for the grid data structure, the Dependency Analysis Module produced space usage metrics for the array and list styles that were deemed "too close" to make a decision. Hence, the execution time metrics were used to make the choice, and the faster array style won. For the next layer of this hierarchically composed data structure, the goal to conserve memory took precedence and the list style, being the clear winner on space usage, was selected.

9.2.2. A Gate Array Routing Task

To test the functionality and correctness of each of the ELF-synthesized routers, we performed a modest routing task with each. The task was to route about 1400 multi-point nets for a 735 gate netlist placed in a 900 gate, gate-array image. This is a synthetic gate array, developed specifically to test routing algorithms in [5], with the structure of a small ALU with a register file, multiplexors and connecting busses. This is the same benchmark used to test annealing-based placement algorithms in [6]. Some performance statistics for these routers are given in Figure 9-3.

Experiment	Space Required (bytes)	Routing Time (VAX 8800 sec.)	Wires Routed
1	14606	87.7	1100/1100
2	6456	132.5	1394/1394
3	8208	115.5	1394/1394
4	9376	59.5	1394/1394

Figure 9-3: Gate Array Experimental Routers: Execution Statistics

Note that Router 1, which was synthesized for speed, did indeed complete the task in less time, about 50% faster that Router 2. However, Router 2, which was synthesized to conserve memory space, required 6K bytes of space compared to Router 1's 14K bytes for the identical problem. Router 3 executed more slowly than Router 1, even though both were tuned for speed. The difference is the change in the expansion phase to handle the decomposed two-point nets. Router 4 ran fastest of all showing that the gain of increasing router cell size (i.e., a coarser routing grid) overcame the time required to expand extra cells for diagonal wires. Finally, and perhaps most critically, all four routers succeeded in routing all of the nets.

9.2.3. Comparison with Hand-crafted Code

Finally, we compare the four gate-array style routers with a hand-crafted version. This version was written in about 1 month for a graduate level VLSI CAD Design Course taught at Carnegie Mellon University. The hand-crafted version is rather more sophisticated (e.g., it uses more sophisticated cost functions, and it is has the additional capability of routing the I/O pads on the gate array). In addition, the hand-crafted version relied on an extensive library of generic list processing routines to simplify the programming task.

Figure 9-4 details the execution time results for this hand-crafted version. Experiments 1 and 2 compare favorably with the hand-crafted version. As expected, the hand-crafted version routed all of the nets. Interestingly, the execution time for the hand-crafted version is slower than than both Experiments 1 and 2. This reflects the use of the library of (apparently slow) generic list processing routines during the expansion phase. The hand-crafted version can not handle the fabrication changes required to meet the specifications for Experiment 3 or Experiment 4. Extensive modification would be required to modify the hand-crafted version to meet the new fabrication requirements, i.e., considerably more time than the 1.5 hours it would require

Automatic Programming Applied to VLSI CAD Software: A Case Study 179

Experiments	Routing Time (VAX 8800 sec.)	Wires Routed
hand-crafted w/ experiment 1,2 requirements	120.4	1394/1394
hand-crafted w/ experiment 3 requirements	Unable to route	X
hand-crafted w/ experiment 4 requirements	Unable to route	X

Figure 9-4: Synthesized Gate Array vs. Hand-crafted Router Comparison

ELF to build these routers from scratch.

9.3. Printed Circuit Board Style Router

For our next set of experiments, we synthesized a global router for printed circuit board applications. Interestingly, this problem applies the ELF synthesis architecture to generate a "real-world" router for a class of DEC VAX mainframe boards. We summarize the basic characteristics of this router below:

- **Router 5:** Ordinary fabrication constraints, e.g., a two-dimensional routing grid of regions covering a multi-layer printer circuit board. There are two routing layers, and each layer is constrained to route in only one direction. Via usage must be minimized and the cost-function must not only match wiring demand to wiring capacity, but take into account via usage when necessary. In other words, each cell in this routing problem is a small region of the board, whose edges have a small, fixed capacity for wires to cross through them, and whose interior has a small, fixed capacity for vias, used when a path changes directions by changing layers. Each of these two capacities must be reflected in the cost of expanding a cell during the search phase of the router. Figure 9-5 shows the different via and wiring cell capacities as wires and vias are routed. In addition, the input netlist and output formats must match the requirements of an operational industrial system. ELF was directed to synthesize this router to optimize its speed.

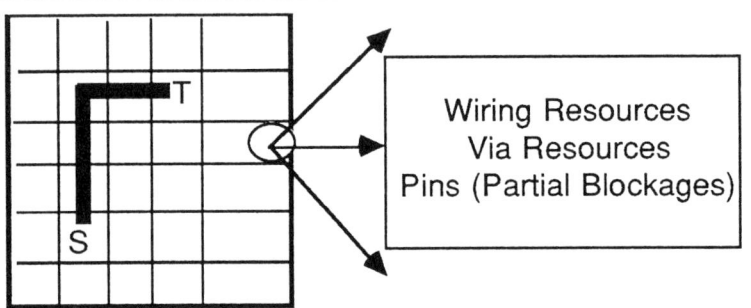

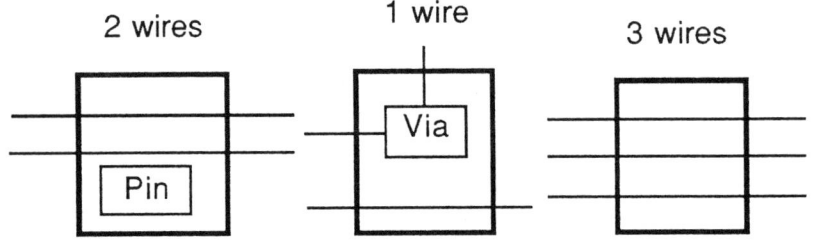

Figure 9-5: Different Via and Wire Traversal Cell Capacities

9.3.1. ELF-Synthesized PCB Router

The PCB router was synthesized in about two hours on a VAX 8800. Figure 9-6 gives the number of rules fired within each stage of the ELF architecture and the resultant number of lines of code.

The proportion of rules fired in each stage is similar to the gate array, yet the resultant code is quite different. Figure 9-7 summarizes the ELF decision process for the VAX mainframe board application. Interestingly, ELF used the identical set of ADL algorithm descriptions used in the gate-array router applications. But the domain knowledge-driven decisions in the Selection Stage reflect the different PCB fabrication requirements. For example, ELF chose a depth-first search, as it did in a few of the gate-array routers. But this application requires the instantiation of the search to take into account the effects and use of vias. The cost metrics used in the selection process reflected the impact of vias on both the size of the *grid* data structure and on the wavefront size. A bounding-box search modification was added to limit the search time.

The cost function, the viability test applied to expand only legal cells during the expansion search, and the underlying grid data structure are modified to reflect vias. Figure 9-8 shows the ELF-designed *grid* data structure for the VAX mainframe board application. This data structure is more complex than in the previous gate array applications to handle the two consumables, wiring and via capacities, for each cell.

9.3.2. A PCB Routing Task

To test the functionality and correctness of this ELF-synthesized PCB router, we obtained two industrial board routing examples from DEC. These two examples have the following characteristics:

- **DEC Easy Example**: This is a modest example selected mainly to insure the correctness of the ELF PCB router. It requires no extreme sophistication in the router to successfully complete this task. This boards has 119 connections, including power and ground connections, in about 25 square inches of PCB area.

	VAX Mainframe Board Experiment
Lines of C Code	1744

ELF Rule Firings
  Input
  Domain Interpreter
  Code Generator
  Internally Generated
	
	72
	7434
	30,443
	3

Total	37,952

Figure 9-6: Code Size and ELF Rule Firings for VAX Mainframe Experiment

Specifications	Vax Mainframe Board
Fabrication Minimize	Printed Circuit Board speed
Decisions	
Search Grid Wavefront	depth-first array of array ordered list

Figure 9-7: VAX Mainframe Board Experiment Decisions

Automatic Programming Applied to VLSI CAD Software: A Case Study 183

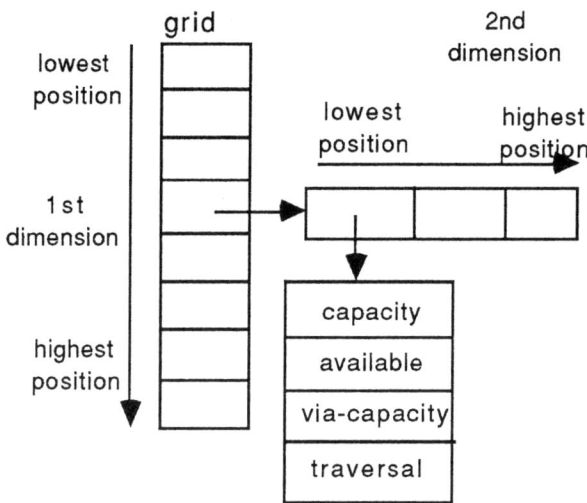

Figure 9-8: VAX Mainframe Board Experiment *grid* Data Structure

- **DEC Difficult Example**: This is a much more interesting example because, for all intents and purposes, it is regarded by experts as being unroutable by automatic tools [4]. In particular, the board is extraordinarily dense, and requires manual interaction to finish its wiring. The fabrication rules must be broken to complete all of the wiring. The boards has about 2900 connections in 111 square inches. This is a particularly appealing example for us, even though we have no real hope of completely routing it, because it has been used inside DEC as a *test* of the quality of externally supplied vendor routers. In this it can be regarded as playing the same role that a number of other well-known "difficult" example problems have played to measure the quality of particular routers, e.g., Deutsch's difficult channel [3], or Burstein's difficult switchbox [2].

Performance statistics for the easy example appear in Figure 9-9. Note that ELF succeeded in routing all of the nets, including power and ground. ELF noted which nets were power and ground nets in the input netlist file and routed those nets first, then followed the shorted-net-first rule for the remainder of the

Experiment	Routing Time (VAX 8800 sec.)	Wires Routed	Wirelength (inches)
5: Easy PCB router	124	119/119	382

Figure 9-9: DEC Easy Example Execution Statistics

nets.

Performance statistics for the difficult example appear the first line in Figure 9-10. Note that ELF, unlike the easy example, did not succeed in embedding all the wires, even though this is a higher-level global routing problem. To put this result in perspective, we must compare to an industrial-quality solution. Figure 9-11 shows the global routing result for the DEC easy problem. Figure 9-12 shows the global routing result for the DEC difficult problem.

9.3.3. Comparison With a Production-Quality Router

No direct comparisons were available for the easy example (essentially because of its age), but routing experts inside DEC, after examing ELF's solution, have noted that it is of comparable quality to internal tools [4]. However this is clearly not a sufficient test since this routing problem is so simple.

Fortunately, direct comparison to a production-quality solution was available for the difficult example. These comparisons are surveyed in Figure 9-10. We note immediately that neither router finishes all connections: ELF embeds 91%

Experiments	Space Required (bytes)	Routing Time (VAX 8800 sec.)	Wires Routed	Wirelength (inches)
5: PCB router	41K	489	2552/2792	3696
DEC Production-Quality Router	2.2MB	512	2594/2792	4090

Figure 9-10: Comparison with a Production-Quality Router

of the nets, while this particular production run finishes 94%. The production solution is only somewhat better, mainly because it has a sophisticated ripup/reroute strategy that ELF currently does not have in its ADL library. However, the DEC solution requires significantly more space since the production system represents all six layers of the printed circuit board, even though only two layers are routable. Both ELF and the production solution used a similar number of vias. We regard this as a very successful result given ELF's current knowledge of routers. ELF produces about the total wirelength expected, given that ELF routed fewer wires.

9.4. Macro-Cell IC Style Router

Our final ELF-synthesized router illustrates the variety of applications targetable by the ELF synthesis architecture. Here, we have generated a router that is based on a substantially different algorithmic approach than used in the previous routers. We summarize the basic characteristics of this experiment below:

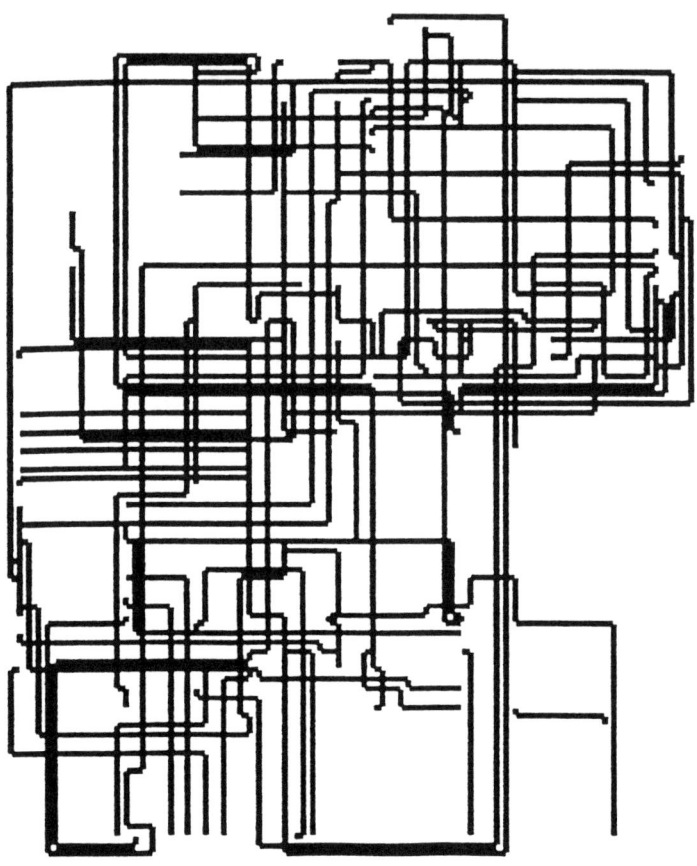

Figure 9-11: Experiment 5: DEC Easy Problem Global Routing Result

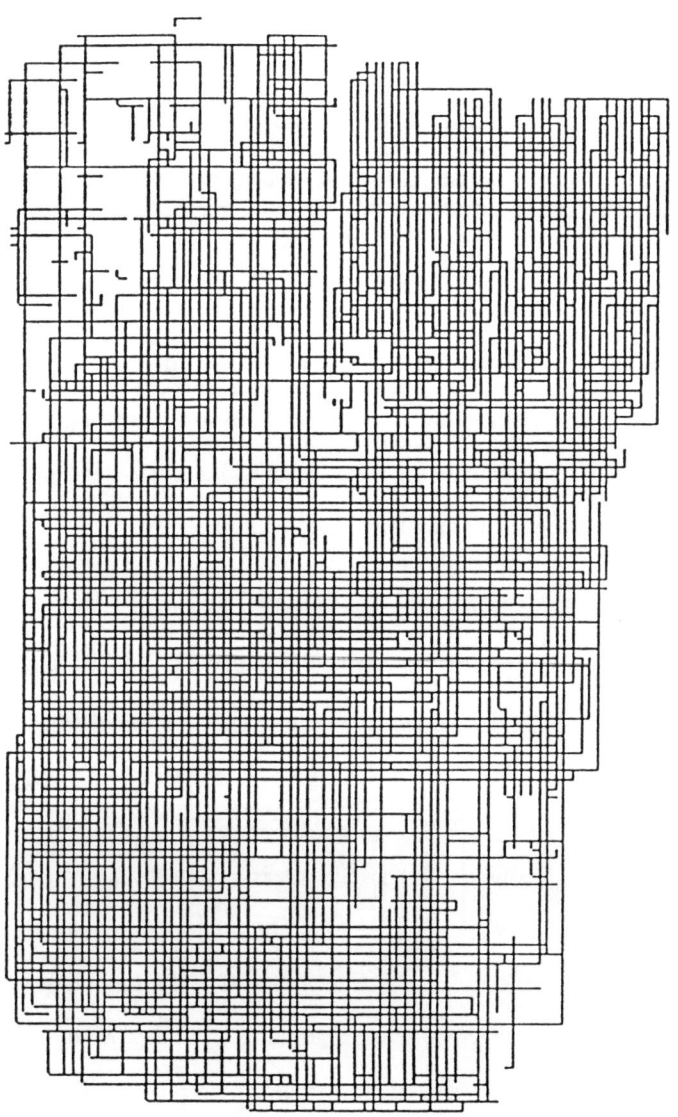

Figure 9-12: Experiment 5: DEC Difficult Problem Global Routing Result

- **Router 6:** Ordinary fabrication constraints for macro-cell IC's. Typical constraints for global routing of such ICs: routing regions are arbitrarily shaped rectangles with capacities for wires to cross their boundaries. Unlike all our previous routers, these regions do not form a simple two-dimensional grid of cells each with four neighbors for Manhattan routing (or eight neighbors if 45° wiring is allowed). Instead, the underlying data structures holding region-to-region adjacency information is a *graph* of these regions [7]. Expansion operations now must traverse the graph to find reachable neighbors, and not simply look at neighbors in a two-dimensional array. This basic difference is illustrated in Figure 9-13.

9.4.1. ELF-Synthesized Macro-Cell IC Global Router

This router was synthesized in about one and one-half hours on a VAX 8800. Figure 9-14 gives the number of rules fired within each stage of the ELF architecture and the resultant number of lines of code.

The proportion of design interaction rules fired to the number of lines of generated code is greater for this experiment than in the grid-based routers. This is due to the change in adjacency definition from the *grid* data structure organization (e.g., *grid* indices) to a new *adjacency definition* data structure that is strongly dependent (within the dependency graph) upon the *grid* data structure. Figure 9-15 summarizes the ELF decision process for the graph-based router. ELF used the identical set of ADL algorithm descriptions used in the previous router applications. ELF retargeted the ADL algorithm descriptions to meet the specifications of graph-based algorithms. In other words, these ADL schemas do indeed specify abstractly the computations required to build a global router, and ELF is capable of refining them, using its transformation mechanism, into a very different router with radically different data structures. In particular, the *grid* data structure is not really a grid or two-dimensional array at all, but is

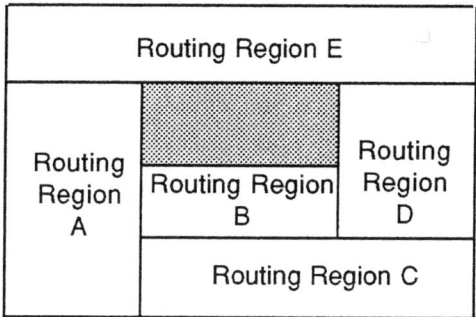

Cell Space Configuration
where each Region Graph is used

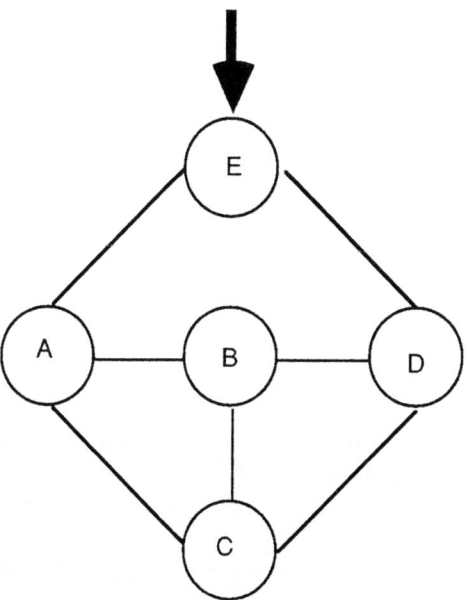

Resultant Graph

Figure 9-13: Comparison Between Grid and Graph Routing Schemes

instead a graph built as a multi-rooted tree of lists, with each list defining the adjacency requirements.

	Graph-based Experiment
Lines of C Code	1527
ELF Rule Firings	
Input	70
Domain Interpreter	6273
Code Generator	24104
Internally Generated	45
Total	30,492

Figure 9-14: ELF Rule Firings for Graph-Based Experiment

Specifications	6
Fabrication	IC
Minimize	Time
Alg-type	Graph
Decisions	
Search	best-first
Grid	array of array
Wavefront	array of list

Figure 9-15: Graph-Based Experiment Decisions

9.4.2. A Macro-Cell IC Routing Task

Experiment	Routing Time (VAX 8800 sec.)	Wires Routed
6	1.4	10/10

Figure 9-16: Graph-Based Execution Statistics

To test the functionality and correctness of the ELF-synthesized router, we performed a very simple synthetic routing task. The intent is just to show the successful operation of the graph-based synthesized router. The number of cells and nets are quite small (5 cells and 6 nets), and as expected, ELF successfully completes this routing task. Figure 9-16 details the performance statistics.

To further show the correct operation of the graph-based synthesized router, we ran two experiments on identical sets of input data. The first set ran under normal channel capacities. The second set ran with several of the channel capacities blocked. A correct graph-based router recognizes the blockages and reroutes those nets that previously crossed that channel to other channels. ELF correctly recognizes and reroutes these nets. Figure 9-17 illustrates the operation of the graph-based synthesized router using the two sets of input data specifications.

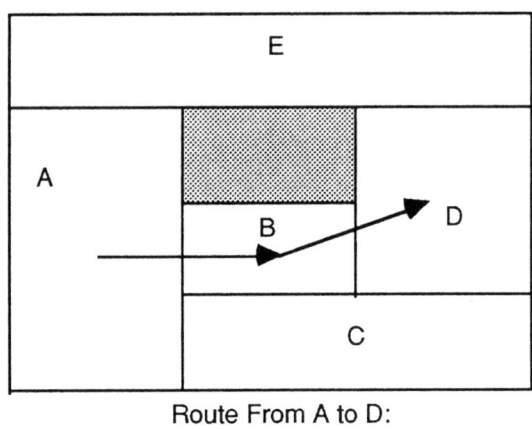

Route From A to D:
A->B->D

Remove Region B:

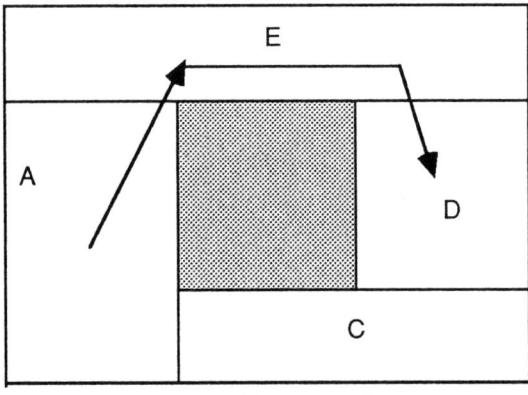

Route From A to D:
A->E->D

Figure 9-17: Effects of Blocking a Channel in a Graph-Based Router

9.5. Chapter Summary

This chapter demonstrates the practicality of the ELF approach. We described six experiments illustrating the wide range of applications, algorithms and fabrication technologies supported by the ELF synthesis architecture, and its current complement of router domain knowledge and ADL algorithm schemas. Most importantly, all six routers performed successfully, either routing all of the test input nets, or in the case of the DEC difficult example by performing up to the standards of a production quality routing system. We regard these experiments as a very satisfactory set of results for the ELF prototype.

References

[1]   A. Aho, J. Hopcroft and J. Ullman, *The Design and Analysis of Computer Algorithms,* Addison Wesley, 1974.

[2]   M. Burstein and R. Pelavin, "Hierarchical Wire Routing", *IEEE Transactions on Computer-Aided Design*, Vol. CAD-2, October 1983.

[3]   D. Deutsch, "A 'Dogleg' Channel Router", *Proceedings 13-th Design Automation Conference*, IEEE, 1976.

[4]   M. Doreau, "Private Correspondence, February 1989".

[5]   R. A. Rutenbar, *A Class of Cellular Computer Architectures To Support Physical Design Automation,* PhD dissertation, University of Michigan, September 1984.

[6]   R. A. Rutenbar, "Simulated Annealing Algorithms: An Overview", *IEEE Circuits and Devices*1989, pp. 19-26.

[7]   J. Soukup, "Circuit Layout", *Proceedings of the IEEE*, Vol. 69, October 1981.

Chapter 10
Conclusion

10.1. Summary

Real-world engineering design tools often exhibit two salient features: they must address a variety of technologies to be viable, and as they grow to accommodate new technologies, they evolve into large, baroque systems. This increasing need to handle new technologies has placed a burden on the design and implementation of VLSI CAD tools: development of adequate tools often lags behind the introduction of new technologies. In many cases, new technologies are unusable for long periods of time due to the failure of current CAD tools to assimilate new technologies in a timely fashion. Upgrading large, baroque software systems is difficult, expensive, and fraught with pitfalls.

In this book, we suggest that for at least some real-world applications, like maze routing software, automatic program synthesis is workable. However, we believe strongly that it is essential to integrate both generic *program synthesis knowledge* and *domain-specific knowledge* to provide quality solutions for real-world problems. Towards this end, we have developed a program synthesis architecture that relies on using these two critical knowledge sources to select appropriate algorithms and data structures, and then transforms these selections into executable code. A prototype system, ELF, working in the domain of IC and PCB wire routing tasks, demonstrates the feasibility of utilizing domain knowledge within a program synthesis architecture.

We believe that ELF has demonstrated the feasibility of applying program synthesis techniques to automatic tool generation. The contributions of this work are as follows:

- **Introduced** a synthesis architecture which integrates the effects of application requirements (domain knowledge) with the actual code requirements (program synthesis knowledge).
  - **Developed** a selection mechanism to choose the appropriate data structure and algorithm schemata. The decision process integrates both *domain* and *program synthesis knowledge* within the selection mechanism. The selection mechanism explicitly acknowledges the interdependent nature of data structure and algorithm selection by iteratively selecting and refining their respective abstract representations.
  - **Developed** a domain knowledge-driven transformation process that explicitly requires both *domain* and *program synthesis knowledge* at each transformation step.
- **Demonstrated** the need for both domain knowledge and program synthesis knowledge within a controlling architectural mechanism. The two knowledge sources are integrated and exploited in real-world synthesis applications. This work demonstrated, using a real-world example, the nature and necessity of controlling the interaction between *domain* and *program synthesis* knowledge in the generation of code from high-level specifications.
- **Demonstrated** that real, workable CAD tools, in particular, routers in several algorithm, application, and fabrication technologies can be synthesized automatically from high-level specifications.
- **Demonstrated** preliminary evidence that a synthesis system can *add* new domain knowledge and then produce new variants of routers using that domain knowledge.

10.2. ELF: Hindsight and Evolution

ELF demonstrates that program synthesis can really work, especially in well-understood, so-called routine design problems like maze routing. But hindsight, and ELF's successes, both suggest several directions for ELF's subsequent evolution.

There are six major areas of future work within the ELF synthesis architecture. First, ELF's domain knowledge is still pretty simple, and one obvious direction for future work is aggressive tuning and enhancements, both to rules and ADL schemas, of ELF's current repertoire of routing knowledge. Second, the ELF synthesis architecture is currently implemented in a possibly outdated language, OPS5. Restructuring the synthesis architecture into another programming language, such as OPS83 [1], or Soar [2], can decrease the design time without eliminating the benefits of a rule-based implementation. However, with the now well-defined synthesis architecture, it is perhaps debatable that a rule-based implementation is necessary for all parts of ELF. In hindsight, we are intrigued with the possibility that ELF could be re-implemented in an algorithmic language (perhaps C or an object-oriented variant of C, such as C++ [3] with a few (possibly large) rule-based components.

Third, the ELF system knowledge base is currently a bit too amorphous. The knowledge operators used in each of the three architectural stages need to be more rigorously defined. Defining the knowledge operators defines the methods by which *domain knowledge* and *program synthesis knowledge* interact. When the methods can be enumerated, the representation of both domain and program synthesis knowledge can be standardized.

The fourth major area requires the standardization of domain and program synthesis knowledge, namely, the development of a front-end knowledge acquisition tool for the domain knowledge base. Development of such a knowledge acquisition tool will allow the retargeting of the ELF synthesis architecture to other application domains.

The fifth major area is the specific targeting of the ELF synthesis architecture to produce production-quality routing software. This will require the development of practical user-interfaces into the Input Stage, ELF error handling formalizations, etc. Modification to the ELF system requires knowledge not only about the routing domain, but also of about half a dozen internal representation entities, i.e., constraints, tasks, ADL, and data structure internal representations. Actual users or maintainers of the ELF system may find that the number of internal representation entities to be too great for efficient maintenance. While it is not within the scope of this book to answer such user-interface and maintenance questions, it is still worth while to suggest that the simplification of these representations should be a primary focus of transferring this work into production form.

The sixth and final major area of work is to optimize the VHLL used as algorithm representations. ADL was developed as a simplified subset of SETL for expedience. In hindsight, the simplification was not necessary and indeed, acts as something of a deterrent to simple algorithm representation. The English-language simplification tended to confuse more than liberate and thus, did not work as well as anticipated. The more powerful SETL-like features of set manipulations without the English-language baggage may represent algorithms more efficiently and correctly.

Finaly, it is worth mentioning that there appear to other applications for which an ELF-like synthesis architecture may be practical. An example of another CAD application to which this approach seems applicable is the design of custom high-level simulators. Such compiled simulators are often a critical part of the verification of many high-level architectures, from mainframe design to the design of real-time systems. Their construction often involves integrating a simulation kernel with custom code for each relevant module of the design; this task is always tedious, often error prone, and, in some sense, routine. Moreover, such simulators are often heavily loaded with custom instrumentation for measuring performance. By abstracting some of the basic structure of these application programs to the level of an ELF-style generator, we may be able to reduce the development time for these routinely-designed, yet sophisticated custom tools.

In the domain of engineering design software, where programs are often exquisitely sensitive to changes in some assumed technology, we believe the idea of domain-specific tool generators will ultimately become a practical approach for managing some important class of software tools.

References

[1] C.L. Forgy, "The OPS83 Report", Tech. report, Carnegie Mellon University, May 1984.

[2] J.E. Laird, A. Newell, and P.S. Rosenbloom, "Soar: An architecture for General Intelligence", *AI Journal*, Vol. 33, September 1987, pp. 1-64.

[3] B. Stroustrup, *The C++ Programming Language,* Addison Wesley, 1987.

Appendix I
Router Specification Manual

I.1. Syntax Description

Input to the ELF generator is semantically arranged in hierarchical levels. Each level contributes information by determining the semantic context of each input. The user directs router synthesis by a sequence of input keywords. This sequence may either be in a file or may be interactive. Most typically, the user will fill in a template of available constraints to describe the functionality of the desired router. A keyword may have one of two functions. The first function is hierarchical management. This keyword organizes the input keyword sequence by determining the context of the following constraints. This is equivalent to going up or down one level of the hierarchy. The second keyword function chooses the actual specifications of the desired router. The semantics of a specification constraint is determined by the context as defined by the input management constraints. While these specification constraints are not inviolate, ELF makes every effort to comply with the constraints as dictated by the user during the course of synthesizing the router.

The syntax of the two keyword functions are different. Input management constraints consist merely of the keyword itself. The keyword *done* is special in that it exists in every context and acts to switch off the most current input context. This is equivalent to going up one level in the hierarchy. Specification constraints consist of the keyword plus a parameter value. For example, the specification constraint in the algorithm context area, *internal_sort yes*, will direct ELF to order the nodes within a given net before attempting to route.

Invalid parameter choices are maintained, but are not considered by the ELF selection mechanism. No invalid parameter notification message is returned to the user. The lack of notification allows for new knowledge acquisition without requiring any modification of the ELF input module. All occurrences of a specification constraint are maintained by ELF. Due to ELF's implementation, only the final occurrence will affect router synthesis [1].

Some keywords, such as *netlist* and *output*, take descriptive file names as parameters. The syntax of these files is given where the appropriate constraint parameter is discussed.

I.2. Constraint Level Structure

This section describes the constraint semantics in each hierarchical level in a top-down manner. All possible parameter options are enumerated. Finally, constraint interaction is explored.

I.2.1. Top-Level Constraint Specifications

At the top level of the hierarchy are the three basic routing constraint areas: algorithm, application, and fabrication. The algorithm constraints give ELF specific guidance on router algorithm selection. The application constraints describe the desired routing task. Finally, the fabrication constraints define the physical environment of the carrier in which the the router operates. All of the top-level constraints are input management constraints. Use of any of the three keywords *algorithm*, *application*, and *fabrication* will put the user into the appropriate constraint area. This is equivalent to moving down the hierarchy. The keyword *done* at the top-level will exit the user from the user constraint input section and begin the ELF synthesis of the desired router.

Each of the following sections enumerate the options available at each hierarchical level within the top-level constraint sections.

I.2.2. Algorithm Constraints

Algorithm constraints pertain to the form of algorithmic attack to be used in a particular routing task. These constraints indicate the user's preference for a particular algorithm over other possible algorithms. Figure I-1 shows the possible constraints and their respective available options. Constraints labeled with an asterisk are required in a template, pound signs indicate numerical responses.

```
algorithm
  net_sorting                 shortest_net_first,
                              longest_net_first, no
  node_sorting                yes, no
  cost_function
    association               cell, boundary, both
    congestion_multiplier     #
    via_multiplier            #
    layer_multiplier          #
    penalty                   linear, exponential, step
  input                       descriptive_netlist_file
  output                      descriptive_output_file
  expansion
    search_phase              depth-first, best-first
    search_restriction        bounding-box, none
    granularity               #
*   minimize                  time, space, wire-length
*   input_net_composition     two-point, multi-point
```

Figure I-1: Algorithm constraints: * required

I.2.2.1. Net_sorting

The *Net_sorting* input constraint allows the user to select any appropriate router input netlist sorting algorithm. Use of this constraint can affect the routing quality, because routing quality is a direct function of the order of net wirings.

The one specifiable parameter here specifies the sorting operation to be used on the input nets before routing. The available options for this parameter are *longest-net-first*, *shortest-net-first* and *no*. The selection of either *longest-net-first* or *shortest-net-first* will direct ELF to attempt to sort the input nets, using a bounding box metric, before submitting the nets to the routing phase. The order of net submission is to route the longest net first for the *longest-net-first* parameter value and vice versa for the *shortest-net_first* parameter value. The selection of *no* will direct ELF to submit the nets in the same order as the input netlist sequence. Any of these parameters may be indicated from combinations of other constraints, for example, the shortest_net_first parameter value may be indicated in congested gate array applications.

I.2.2.2. Node_sorting

The *node_sorting* input constraint allows the user to select any appropriate input node sorting algorithms. Use of this constraint can affect the routing quality, because routing quality is a direct function of the order of node-pair orderings.

The one specifiable parameter here specifies the sorting operation to be used on the input node-pairs within each of the input nets before routing. The available options for this parameter are *yes* and *no*. The selection *yes* will direct ELF to attempt to sort the nodes of the input nets using a minimum spanning tree algorithm. Knowledge of the minimum spanning tree algorithm is embedded within ELF. The actual order of the nets themselves are not affected. The selection of *no* will direct ELF to submit the nodes within the nets in the same order as in the input netlist sequence.

ELF may internally generates an node sort if the fabrication constraints restrict connections to occur at node points only, then a minimum spanning tree ordering of the nodes will alleviate congestion.

I.2.2.3. Cost_function

The *cost_function* input constraint allows the user to customize the cost function to improve routability, and to override non-essential fabrication constraints. The cost function can be customized to reflect congestion concerns, excessive via-usage, and non-conforming layer direction wire layout. The arithmetic cost function base type may also vary, e.g., linear, exponential, etc.. The cost function determines the ordering of cell expansion with the expansion search task. Congestion is an accurate model for a cost function and is defined as cell wiring capacity divided by cell wiring availability for cell traversal, but is defined as the difference between the cell wiring availability number for the reached cell and the expanded cell for crossing cell boundaries. ELF generates a cost function using these parameters and the appropriate congestion definition using the mathematical function shown in Figure I-2.

The following constraints are available in the cost_function definition area.

Cost function =

Congestion_multiplier * (capacity/available) +
Via_multiplier * (via capacity/ via available) +
Congestion_multiplier *
(reached cell capacity/reached cell available -
expanded cell capacity/expanded cell available)

Figure I-2: Cost Function Determination

Association

The *association* input constraint allows the user to specify cost_function components for alternative costing strategies, such as cell traversal and the crossing of cell boundaries. How this cost function is bound to the process of traversing a cell affects how cell are reached.

The one specifiable parameter here defines where the cost function is to be bound. The possibilities are the expanding *cell* traversal, the *boundary* between cells, *both*, or *none*. The *both* option results in a more complicated, though more realistic, cost function.

Congestion_multiplier

The *congestion_multiplier* input constraint allows the user to weight the contribution of the association portion, defined above, of the cost function. A more realistic cost function can result from appropriate weightings. The parameter is a numeric multiplier value.

Via_placement

The *via_placement* input constraint allows the user to account for via usage in the cost_function. This is an important variable in PCB applications where via placement and usage are at a premium. Vias can be placed according to fabrication constraints or they can be allowed to freely place themselves on the routing surface. While vias eventually must meet fabrication constraints, allowing free via placement gives the user an upper bound on via requirements.

The one specifiable parameter defines from where the via placement rules function is to be chosen. The possibilities are from *fabrication* rules or *none*. The *fabrication* option interacts with the *vias* constraint available in the PCB constraint area to derive a percentage available via count for the desired application.

Via_multiplier

The *via_multiplier* input constraint allows the user to weight the contribution of the via usage portion of the cost function. A more realistic cost function can result from appropriate weightings. The parameter is a numeric multiplier value.

Layer_component

The *layer_component* input constraint allows the user to deviate from the preferred layer directionality. The added flexibility of going against the preferred direction for a layer can improve routability.

The one specifiable parameter defines from where the directionality rules function is to be chosen. The possibilities are from the *direction* rules or *none*. The *direction* option interacts with the *layer* constraint in the PCB or IC constraint areas to derive the appropriate layer variable implementation.

Layer_multiplier

The *layer_multiplier* input constraint allows the user to weight the contribution of going with and/or against the direction rules. A more realistic cost function can result from appropriate weightings. The parameter is a numeric multiplier value.

Penalty

The *penalty* input constraint allows the user to choose the type of function to be used as the base cost_function. The options available for the parameter are: *linear* and *exponential*. The base function of each component of the cost_function is specified by this constraint. For example:

- **linear**: if the user specifies a linear cost function applies the cost function to cell boundaries, with a congestion multiplier of 10, ELF may generate (depending on other application and fabrication constraints, in this a gridded global gate array router) the following cost function: cost = 10 * (capacity/available of reached cell - capacity/available of expanded cell).

- **exponential**: if the user specifies an exponential cost function and applies the cost function only to cell traversal, with a congestion multiplier of 10, ELF may generate (depending on other application and fabrication constraints, in this case a gridded global gate array router) the following cost function: cost = 10 * exp (capacity/available).

I.2.2.4. Netlist

The *netlist* constraint specifies the composition of the input netlist. The one parameter has two possible options: either the name of a description net list file or *none*. The syntax of the input netlist description must be in the following syntactic format. This format describes the composition of the input netlist. Parameters in this description define a model of the desired input (or output) netlist. The BNF-like format has two sections. The first section, the **define** section, tailors the netlist input task to the actual numerical parameters present in the input netlist file. For example: each record in the input file may be identified by a set of keys. The input file net records may be identified by specific key values. These values must be defined in the **define** section. The second section, the **model** section, describes the functionality of the input file in a BNF-like format. The BNF synthesis machine contains knowledge of the following keywords specific to the routing domain. They are: FILE, NET_NUMBER, NODE_POSITION *, and +. The keyword FILE is the final BNF result and is therefore not used to synthesize the input routine. The keyword NET_NUMBER refers to the net numbering scheme used by the input netlist format. The following keywords, with their definitions, are available.

- Number_of_pads: The number of pads to be included in routing.
- Pad_number: The specific identification number of the pad to be routed.
- Pad_position: The location of a pad in the coordinate frame.
- Number_of_nets: The number of nets to be routed, and are included in the input netlist.
- Net_number: The specific identification number of the net to be routed.
- Number_of_nodes: The number of cells to be routed, and are included in the input netlist.
- Number_of_inputs: The number of nets on the cell being defined.
- Node_number: The specific identification number of the cell to be routed.
- Nodes_on_net: The number of cells on the net being defined.
- Node_position: The location of a cell in the coordinate frame.
- Pin_number: The specific connection pin within the cell being routed.
- End_marker: An end-of-input marker, defaulted by ELF to be -1^+.

The symbols * and + have their standard BNF meaning of *none or more* for the * symbol and *at least one* for the + symbol. The keywords, Net_number and Cell_number, and special symbols can be intertwined to define the appropriate netlist input file composition. If the option *none* is chosen, then the default composition given in figure I-3, in keyword format, is used. In this example, the number of cells to be routed is first defined. Each succeeding line consisting of a cell definition, including the number of nets of the particular cell, the net identification of each net of this cell, and finally the cell position. In this way, each cell, its position and connectivity, is defined.

```
DEFINE
MODEL
    file <= FIRSTRECORD RECORD
    FIRSTRECORD <= number_of_cells
    RECORD <= RECORD /
              CELL_NUMBER INPUT_NUMBER
              NET_NUMBER + NODE_POSITION /
              EOF
END
```

Figure I-3: Input Netlist Definition Example

I.2.2.5. Output

The *output* constraint specifies whether an output routing list of wiring paths is required as well as its composition. The one parameter has two options, the name of a descriptive list file or *none*. The syntax of the output list description must be in the following syntactic format. This format describes the composition of the output routing list. The following keywords, with their definitions, are available.

- Number_of_nets: The number of nets routed.
- Net_number: The specific identification number of the routed net.

- Number_of_nodes: The number of cells on the routing list.
- Node_number: The specific identification number of the routed cell.
- Nodes_on_net: The number of cells on the routed net.
- Node_position: The physical location of the routed carrier.
- End_marker: An end-of-output marker, defaulted by ELF to -1$^+$.

The symbols * and $^+$ have their standard BNF meaning of *none or more* for the * symbol and *at least one* for the $^+$ symbol. The keywords, net_number and cell_number, and special symbols can be composed to define the appropriate netlist input file. If the option *none* is chosen, then the default composition given in figure I-4, in keyword format, is used. In this example, the number of routed nets is first defined. Each succeeding line consisting of a net identification, then all cell positions through which the carrier was routed.

```
Number-of-Nets
Net_number Node_position +
*
```

Figure I-4: Output Routing List Example

I.2.2.6. Expansion

The *expansion* input constraint area allows the user to select the algorithm used in the expansion search phase of a maze router, restrict this algorithm's search space, and add several quantitative measurements to guide the search. Use of this constraint affects the time necessary to complete the routing task as well as the overall routing quality. The following constraints are available in the expansion search definition area.

Search_phase

The *search_phase* constraint allows the user to choose the desired search algorithm to be performed in the expansion phase of routing. This parameter, in conjunction with the speed/space minimization constraint, is the strongest data structure selection implicant. Knowledge of the search algorithms is embedded within ELF.

The options available are *depth_first* and *best_first*. The first option, *depth-first*, selects a depth-first expansion search algorithm. The cell closest to the target is chosen for expansion. This option does make use of source/target relative placement in the search. The second option, *best-first*, performs the search by expanding the cheapest cells in order. This option does not make use of source/target relative placement.

ELF may internally generate a search phase selection under the following conditions:
- If seeking to minimize speed and one of the search possibilities has a lesser effect on speed cost, given the known and possible data structure selections, then choose it.
- If seeking to minimize space requirements and one of the search possibilities has a lower space requirement and it has a lower affect on the current space requirements, given the known and possible data structure selections, then choose it.

Search_restriction

The *search_restriction* input constraint allows the user to restrict the search space during the expansion phase of routing. This constraint has two parameters. The first has two options: *Bounded-box*, and *none*. The *Bounded-box* option restricts the expansion search space to x% greater than a bounding-box around the nodes of a net, where x is defined in the second parameter. This greatly limits the size of search problem. The *none* option does not restrict the expansion search space. Typical values for the second parameter might be 10, or 20. The bounded-box parameter value may be indicated by combinations of other constraint values, e.g., a congested PCB.

Minimize

The *minimize* input constraint directs the search to minimize either one of the two possible options, *time* or *space*. Expansion search can be streamlined to minimize either one of these two parameters with repercussions being felt in the router implementation.

I.2.2.7. Net_composition

The *net_composition* input constraint specifies the composition of the routing nets during the setup phase. The available options are *two-point* and *multi-point* definitions. ELF synthesizes code to massage the input netlist to accommodate the desired routing capability given in the *routing* constraint. Specifically, the multi-point nets may be broken into a series of two-point nets using the node-sorting algorithm choice chosen by the *Internal_sort* constraint.

I.2.2.8. Routing_composition

The *routing_composition* input constraint specifies the composition of the routing nets during the expansion phase. The available options are *two-point* and *multi-point* definitions. Both options are currently supported. This constraint works in tandem with the *net* constraint to handle setup phase to expansion phase net list incompatibilities.

I.2.3. Application Constraints

Application constraints pertain to the differing task requirements of the router itself. Use of these constraints aid ELF synthesis by describing the routing task requirements. These constraints describe the user's perception of the routing task. Figure I-5 shows the possible constraints and their respective available options once the user has entered the application constraint area.

```
application
* type                    global, detailed
  alg_type                gridded, graph
  sub_type                gate-array, general
```
Figure I-5: Application constraints: * required

I.2.3.1. Type

The *type* input constraint allows the user to select the desired type of maze router to be generated. Use of this constraint greatly restricts the applicability of the synthesized router. This constraint interacts with many of the algorithm constraints in the formulation of the desired router.

Available options are *global* and *detailed*. The first option, *global*, forces the generation of a global router. In this case, the actual physical placement of the wiring paths is not a concern. This first-pass router is concerned with wiring through high-level regions. The output is a listing of the appropriate regions though which a wiring path crosses. The second option, *detailed*, chooses the generation of a detailed router. In this case, the output is the actual wiring location of the connective wiring paths. This option is not currently supported in this version of the ELF implementation.

I.2.3.2. Sub_type

The *sub_type* input constraints further divides the desired application into two categories. The available options are *gate-array* and *general*. The *gate-array* option implies certain routing constraints such as non-routability of those areas taken up by cells as well as physical compatibility with certain implementation choices. The *general* option does not take advantage of these assumptions.

I.2.3.3. Alg_type

The *alg_type* input constraints further divides the desired application algorithm into two categories. The available options are *grid* and *graph*. The *grid* option forces the maze router application onto a grid, thereby implicitly defining node adjacency within the grid constraints, before preforming the expansion search. The *graph* defines each node of each node as a node in a graph network that defines node adjacency.

I.2.3.4. Number_of_nets

The *number_of_nets* constraint informs the router synthesizer of how many nets can be expected to be handled by the router. The default for this number and the following constraint, the *number_of_cells_per_net* is computed to give 100% of the available capacity. This number is not hard-wired into the router, but is used to compute the congestion numbers used in the ELF synthesis architecture decision process.

I.2.3.5. Number_of_cells_per_net

The *number_of_cells_per_net* constraint informs the router synthesizer of how many cells typically comprise a single net. The default for this number and the preceding constraint, the *number_of_nets* is computed to give 100% of the available capacity. This number is not hard-wired into the router, but is used to compute the congestion numbers used in the ELF synthesis architecture decision process.

I.2.4. Fabrication Constraints

Fabrication constraints describe the operating physical specifications of the routing task. Figure I-6 shows the possible constraints and their respective available options once the user has entered the Fabrication constraint area.

```
fabrication
*  units                    metric, lambda
*  pads                     yes, no
*  connections              spanning, steiner tree
*  technology               pcb, ic
*     xcapacity             #
*     ycapacity             #
*     zcapacity             #
*     pcb
*        number_of_layers        #
*        available_positioning fixed, random
*        layer
*           id                   #
*           expansion_direction x, y, z, 45
*           wire_width           #
*           wire_to_wire_space  #
*           via_spacing          #
*           wire_to_via_space   #
```

Figure I-6: Fabrication Constraints

I.2.4.1. Units

The *units* input constraint allows the user to specify the type of units being used in the fabrication specification. The available choices are *english* and *metric*. If *english* is specified, then all fabrication numbers are in mils. If *metric* is specified, then all fabrication numbers are in millimeters.

I.2.4.2. Pads

The *pads* input constraint allows the user to indicate the presence of pads within the routing task. Use of this constraint can allow the generation of the appropriate support of pads in the router. There are two alternatives; either *yes* or *no* are options. This constraint is not currently supported within the current version of ELF.

I.2.4.3. Connections

The *connections* input constraint allows the user to indicate the fabrication process constraints on interconnections. The two options are *spanning* and *none*. This constraint works with the *Internal_sort* constraint to match algorithm requirements, such as expansion search connectivity, with fabrication constraints. A spanning tree interconnection requirement implies that all net interconnects must be at net endpoints (e.g., wire-wrap board applications). Otherwise ELF assume that all current net traversal points are eligible interconnect points.

I.2.4.4. Xsize,Ysize,Zsize

The dimensions of the routing area are given in the *Xsize*, *Ysize*, and *Zsize* input constraints. Each constraints takes a numerical parameter determining the length in whatever units are specified by the *units* constraint.

I.2.4.5. Technology

The *technology* input constraint chooses which of the two possible fabrication processes to target the routing task. The options are *PCB* and *IC*. Constraints given to the non-selected fabrication process are ignored.

I.2.4.6. PCB

This constraint, *pcb*, selects a printed-circuit-board fabrication process. It is a hierarchical input management constraint. The following constraints are processed relative to a printed-circuit-board fabrication specification. Issues peculiar to PCB's such as routing layer directional capabilities and actual board fabrication parameters with respect to via placement affect routing quality.

I.2.4.7. Number_of_layers

The *number_of_layers* input constraint specifies the total number of routing layers available in the PCB technology. The user specifies the number of layers as a parameter to the constraint.

I.2.4.8. Available_via_positioning

The *available_via_positioning* input constraint describes the available positioning of vias available in the PCB technology. The user first specifies the type of availability, i.e., *unrestricted*, and *fixed*. This constraint interacts with the cost function implementation constraints when forming the desired cost function. *Unrestricted* allows via placement to occur throughout the physical routing domain, subject to via allocation per routing cell. *Fixed* only allows via allocation at specific routing cells.

I.2.4.9. Layer

The *layer* input constraint area allows the user to specify various parameters that can vary per layer such as the layer identification, allowable expansion directions, allowable deviations from the expansion directions, wire width, wire-to-wire spacing, via-width, and wire-to-via spacing. The following constraints are available in the cost_function definition area. This constraint is especially useful in detailed routing applications. ELF currently can not synthesize automatically detailed routing applications.

ID

The *id* constraint binds the following layer constraints to this identification parameter. This parameter remains constant until another *Id* constraint is given.

PCB_expansion_direction

The *pcb_expansion_direction* constraint defines the allowable expansion restrictions. The number of parameters may vary according to the number of allowable expansion directions on the binding layer. This constraints may be any or up to all of the following options: *x, y,* and *z*. This constraint affects capacity initialization in global routing applications.

Violations

The *violations* constraint defines the additional cost of violating the allowable expansion direction constraints for the current layer. This constraint works with the cost_function constraint to develop the appropriate cost_function for the routing task. The allowable options for the one specifiable parameter are *yes* and *no*. The option *no*, for the case of a fabrication technology that does not allow the expansion restrictions to be violated, does not affect the cost_function. The option *yes*, in which case fabrication does allow violation of expansion restrictions does affect the cost_function.

Wire_width

The constraint *wire_width* defines the width of the carrier on the current layer.

Wire_to_wire_spacing

The constraint *wire_to_wire_spacing* defines the carrier-to-carrier spacing on the current layer.

Via_width

The constraint *via-width* defines the minimum width of vias.

Wire_to_via_width

The constraint *wire_to_via_width* defines the minimum space requirements between any wire and any placed via.

I.2.4.10. IC

This constraint, *ic*, selects an integrated-circuit fabrication. It is a hierarchical input management constraint. The following constraints are relative to an integrated-circuit fabrication specification. Issues peculiar to IC's such as routing layer makeup and physical capabilities affect routing quality. Similiar constraints exists in the printed-circuit-board fabrication context, but the semantics and available options are different.

The identical constraints available for the PCB fabrication option are available for the IC fabrication option.

Figure I-7 shows the complete input template options.

```
algorithm
  net_sorting                        shortest_net_first,
                                     longest_net_first, no
  node_sorting                       yes, no
  cost_function
    association                      cell, boundary, both
    congestion_multiplier            #
    via_multiplier                   #
    layer_multiplier                 #
    penalty                          linear, exponential, step
  input                              descriptive_netlist_file
  output                             descriptive_output_file
  expansion
    search_phase                     depth-first, best-first
    search_restriction               bounding-box, none
    granularity                      #
*   minimize                         time, space, wire-length
*   input_net_composition            two-point, multi-point
application
*  type                              global, detailed
   alg_type                          gridded, graph
   sub_type                          gate-array, general
fabrication
*  units                             metric, lambda
*  pads                              yes, no
*  connections                       spanning, steiner tree
*  technology                        pcb, ic
*    xcapacity                       #
*    ycapacity                       #
*    zcapacity                       #
*    pcb
*      number_of_layers              #
*      available_positioning         fixed, random
*      layer
*        id                          #
*        expansion_direction         x, y, z, 45
*        wire_width                  #
*        wire_to_wire_space          #
*        via_spacing                 #
*        wire_to_via_space           #
```

Figure I-7: User Specification Template (* required)

I.3. Input constraint Schemes

We illustrate in this section a few complete specifications to give a sense of the variety of routers supported by ELF.

The first example, shown in Figure I-8, describes a global router for a gate array application. In this example, the user wishes to create a rather ordinary global router. No sorting of the nets is required, a breadth-first expansion is sufficient, and the first path found on expansion will cause the backtrace phase to begin. An output list of every region in each wiring path is required. Only two-point nets are submitted for routing and the router need only handle two-point nets. The target fabrication is for an IC with two layers, one poly and one metal. One layer is constricted to x direction routing only and the other is constricted to y direction routing only.

What should be noted in this example is that most of the parameters could be defaulted but for the sake of completeness, each constraint with its appropriate parameters are provided. The constraints are grouped to indicate constraint semantics. It is anticipated that this template approach to input will be appropriate.

The second example, shown in Figure I-9, describes a detailed 4 layer PCB router. No sorting of the input nets is required, but due to time constraints, a best-first expansion search is preferred and the first path found starts the backtrace phase. An output routing path is required. Both the input nets and the router itself must be capable of handling multi-point nets to improve congestion. The outermost layers are directionally unrestricted, while one of the inner layers is restricted to the y direction and the other is restricted to the x direction. Vias pass through all the layers, are restricted to fabrication requirements and connect to each layer. Note the similarity of the two input examples regardless of the widely different specifications.

```
Algorithm
     expansion
          minimize space
     done
     netlist i.netlist.model
     output o.netlist.model
Application
     Type          global
     Sub_type      gate-array
     number_of_nets 100
     number_of_cells_per_net 3
Fabrication
     Pads          no
     IC
          number_of_layers   2
          Layer
               id 1
               ic_expansion_direction x y
          done
          Layer
               id 1
               ic_expansion_direction x y
          done
     done
```

Figure I-8: Example 1: Global gate-array

```
Algorithm
    Output                      o.netlist.model
    net_composition             multi-point
    routing_composition         multi-point
Application
    Type            detailed
Fabrication
    Pads        no
    PCB
        Number_of_layers   4
        Layer
            ID
            PCB_expansion_direction x y z
        done
        Layer
            ID 2
            PCB_expansion_direction y z
        done
        Layer
            ID 3
            PCB_expansion_direction x z
        done
        Layer
            ID 4
            PCB_expansion_direction x y z
        done
        Vias    fixed
    done
```

Figure I-9: Example 2: Detailed four layer PCB gate-array router

References

[1] L. Brownston, R. Farrell, E. Kant, and N. Martin, *Programming Expert Systems in OPS5,* Addison-Wesley, 1985.

Index

ADL 47, 105, 141
Algorithm constraints 27
    Backtrace optimization 28
    Cell Reentrancy 27
    Detour limits 28, 119, 181
    Expansion path definition 27
    Expansion search selection 27
    Net ordering 27
    Wavefront data structure 21, 28
Algorithm Designer Module 84, 115
    Algorithm representation 116
    Operation 115, 118
    Types of knowledge used 121
    Use of Dependency Analysis metrics 120
    Use of Router domain knowledge 121
Algorithm Schema Representation
    See also ADL
Algorithm stress knowledge 60
Application constraints 26
    Net definition 13, 26, 175
    Router classification 26
Application language syntactic knowledge 90
Automatic programming 33
    Application-generators 37
    Composition-based systems 34
    Generation-based 44
    Generation-based systems 35
    Knowledge characterization 6
    Language-based systems 35
    Transformation-based systems 38

Candidate set 84
Code generation stage 66
Code Generator Stage 127
    Bounding box effects 135
    Diagonal wiring example 138
    Generation of expansion sites 139
    Rule types 137
    Transformation Process 141, 146, 148
    Transformation Process Operation 142
    Transformation redefinition 135

232   Index

    Transformation redirection   135
    Types of knowledge used   127
    Use of Wiring angle definition   147, 148
Code Representation   53
Control Knowledge   62, 111
Cost function   181
    Effects of vias on   181

Data Structure Designer Module   84, 96
    Automatic variable creation   114, 144
    Building interdependency graph   106
    Data structure representation   99
    Data structure styles   102
    Field inversion   112
    Interdependency graph   97, 112
    Operation   96, 104
    Set selection process   104
    Types of knowledge used   104, 97
    Use of Control knowledge   111
    Use of Design interaction knowledge   109
    Use of Program synthesis knowledge   109
    Use of Router domain knowledge   107
    Use of Dependency Analysis metrics   102, 176, 181
Data structure implementation knowledge   83, 90, 92, 93, 104
Data Structure Representation   50
Dependency Analysis Module   84, 90
    Computation of costs   90, 92
    Operation   90
    Set relationships   94
    Types of knowledge used   90
Design generation knowledge   56, 127, 134
Design interaction   153
Design interaction knowledge   83, 90, 92, 93, 97, 104, 109, 127, 134, 138, 140, 164, 165
Diagonal wiring effects   159
Dimension   100
Domain interaction knowledge   60
Domain Knowledge   37
Domain-specific knowledge   7
DRACO project   37

ELF   44, 55
ELF synthesis architecture   45
    Types of input constraints   71, 170
    Validation metrics   171

Fabrication constraints   25
    Multiple routing layers   25
    Preferred routing direction   25
    Preferred wiring direction   180
    Vias   14, 25, 180
    Wiring angle   26, 109, 140, 175
Field inversion   112

Gate array experiments   172

Decision process 173
Execution comparisons 177, 178
Generic syntactic knowledge 7
Graph-based routers 161
Grid data structure 93, 106, 107, 132, 139, 140, 146, 147, 176, 188
   Graph-based implementation 147, 188

Input stage 62, 71
   Inference mechanism 71
   Input template 72
   Operation 71
   Rule types 77
   Template Organization 74
   Types of knowledge used 71
   Use of router dependency knowledge 78
   Use of routing phase requirement knowledge 79

Knowledge Representation 54, 55
   Control 62
   Data structure 59
   Design generation 56
   Domain interaction 60
   Program synthesis 58
   Router dependency 57
   Router structure 56
   Routing phase requirement 57
   Syntactic 59
   See also Specific type of knowledge

Lee-Moore Routers
   See also Routers, Maze

Macro-Cell IC experiment 185
   Decision process 188
   Performance statistics 191

Nets
   See also Application constraints, Net definition

Object-oriented Techniques 34
OPS5 55

Printed circuit board experiments 179
   Decision process 181
   Execution comparisons 184
   Performance statistics 184
Program synthesis knowledge 153, 58, 85, 93, 97, 104, 109, 127, 141, 142, 145, 149, 154, 155, 157, 165
PSI/SYN project 38

Refinement definition 83
Rotuer domain knowledge 154
Router dependency knowledge 57, 71, 78, 79
Router domain knowledge 100, 104, 140, 155, 85, 87, 92, 93, 94, 95, 97, 104, 107,

112, 114, 119, 128, 134, 135, 140, 144, 147, 153, 155, 157, 158, 162, 165, 127
   See also Knowledge Representation
Router phase requirement knowledge 83
Router structure knowledge 56, 72, 78, 83
Routers 13
   Channel 16
   Detailed 18
   Global 17
   Maze 16, 19
   One-Wire-at-a-time 14
   Phases of Maze 19
   Restricted-area 16
   Switchbox 16
Routing cell 21
Routing grid 20
Routing phase requirement knowledge 57
Routing structure knowledge 79

Selection Control Module 84, 86
   Operation 89
   Types of knowledge used 88
Selection definition 83
Selection stage 64, 83
   Candidate set 84
   Selection process 64, 65
   Types of knowledge used 83, 85
     See also Algorithm Designer Module, Data Structure Designer Module, Dependency Analysis Module, Selection Control Module
SETL 35, 44, 47
Software reusability
   See also Automatic programming

Tools
   Custom special-purpose 2
   General-purpose 1

Wavefront data structure 107, 137, 144
   See also Algorithm constraints, Wavefront data structure